Seaweeds for Food and Industrial Applications

Seaweeds for Food and Industrial Applications

Bruno Augusto Amato Borges

Delve
PUBLISHING

www.delvepublishing.com

Seaweeds for Food and Industrial Applications

Bruno Augusto Amato Borges

Delve Publishing
224 Shoreacres Road
Burlington, ON L7L 2H2
Canada
www.delvepublishing.com
Email: orders@arclereducation.com

ISBN: 978-1-77407-797-9 (Hardcover)

ABOUT THE AUTHOR

Bruno Augusto is Aquaculture Engineer and Master on Aquaculture and Fishery Recources from Federal University of Santa Catarina (UFSC). His focus is on writing projects and development of new technologies in aquaculture. He is currently project service provider, consultant and develops research on Biofloc Technology (BFT) cultivation for fish and shrimp.

TABLE OF CONTENTS

LIST OF FIGURES

LIST OF TABLES

LIST OF ABBREVIATIONS

ABA	abscisic acid
CAT	catalase
CKs	cytokinins
CO_2	carbon dioxide
COX	cyclooxygenase
DCCI	Dubai Chamber of Commerce and Industry
DPPH	2,2-diphenyl-1-picrylhydrazyl
DW	dry weight
ESR	electron spins resonance
EU	European Union
FAO	Food and Agriculture Organization
FW	fresh weight
GA3	gibberellic acid
GLA	gamma linolenic corrosive
GRAS	generally recognized as safe
GSH	glutathiones
HA	hyaluronic acid
HPt	histidine phosphotransfer
IAA	indole acetic acid
IMTA	integrated multi-trophic aquaculture
iNOS	inhibited inducible nitrogen oxide
JECFA	Joint FAO/WHO Expert Committee on Food Additives
LCA	life cycle assessment
LCIA	life cycle impact assessment
LF	liquid fertilizers
LSF	liquid seaweed fertilizers
MAE	microwave-assisted extraction
MF	microfiltration

Mt	million tons
NDBP	near-future, deployable biorefinery provinces
NF	nanofiltration
NO	nitric oxide
NPP	net primary productivity
PAR	photosynthetically active radiation
PEF	pulsed electric field
PGE2	prostaglandin E2
PNG	Philippines Natural Grade
PUFAs	polyunsaturated fatty acids
RO	reverse osmosis
ROS	reactive oxygen species
RRs	response regulators
SFE	supercritical liquid extraction
SLF	seaweed liquid fertilizer
SOD	superoxide dismutase
SWE	subcritical water extraction
UAE	ultrasound-assisted extraction
UCP1	uncoupling protein
UF	ultrafiltration

PREFACE

Seaweeds have become an important and integral part of the marine ecosystem and the human exploration for their use and employment in various sectors. The seaweeds provide nutrition in various forms and raw materials for several products.

The world is looking up to the seaweeds as a renewable source of energy and as a source for the security of the food requirements of the people around the globe. Various scientists and researchers have directed their researches in the direction of exploring more about the seaweeds and trying to decipher new ways in which it can be utilized with optimized efforts and the sources it can replace as renewable source of energy.

The importance of seaweeds has taken a great rise in the recent times which is reflected in the way the people are driven by the studies on the subject. This book tries to make the readers aware about the various aspects of seaweeds in the current times and the various kinds of discoveries that have been made in that regard.

The book starts off by introducing seaweeds to the readers along with the historical background of the seaweeds. The readers come to terms with the kinds of seaweeds and the various seaweeds that are commonly found in various parts of the world or are cultured in those parts for industrial and personal use.

The second chapter throws light on the various sources of seaweeds so that the readers are accustomed with the places and the natural forms in which the seaweeds are found. This chapter also throws light on the cultivation techniques and various aspects of handling the seaweeds before, during, and after the harvest time.

The third chapter takes the readers through the nutritional composition of the seaweeds so that they are well-versed with the various health benefits that the seaweeds can provide and the purpose of them being used in the food as well as the health industry.

The fourth chapter throws light on the various ways in which the seaweeds can be used in the food industry or food preparation. The chapter also discusses the ways in which the food application of the seaweeds can have an influence on the health of the people.

The fifth chapter shows the readers the manner in which the seaweeds can be used in the agricultural area as in the form of fertilizers, soil conditioners, nutrient for plants, and other important relatable applications, including the ones that focus on organic farming.

The sixth chapter takes the readers through the various kinds of cosmetic and medicinal applications of the seaweeds by explaining them the properties of the seaweeds that are helpful in medicinal application or the characteristics that enable their application in the cosmetic products.

Seventh chapter takes the readers through the various kinds of challenges that are faced during the extraction of seaweeds from the seas and how the people deal with those challenges. The challenges involve access rights, price, food safety, and other important aspects of usage of seaweeds.

The last chapter takes the readers through the various kinds of market trends that are found in the seaweed products and the manner in which seaweeds are expected to exist and be useful in the future. The chapter tries to convince the readers about the potential uses of seaweeds and how it can be a great source for various kinds of requirements in the coming future.

I hope that this book can be of good use to the readers for the collection of knowledge and important insights into the seaweeds. I also hope that this book can contribute for the knowledge and information requirements in the field of application of seaweeds.

CHAPTER 1

Introduction and Historical Background to Seaweeds

CONTENTS

Seaweeds are becoming an important part of the world that we live in. They have become the source for various products in the recent years and are being looked upon as the source for more products as well. This chapter introduces the readers to the seaweeds and makes them aware about the various types and certain aspects of them in the current times.

1.1. INTRODUCTION

The seaweed (or the benthic marine algae) can be considered as a range of plants that grow up and are found in the marine ecosystems or in the brackish waters. Similar to the plants that are found on the land, the seaweeds comprise of the photosynthetic pigments, which have the ability to produce foods with the help of the sunlight and the nutrient they receive being in the seawater.

The seaweeds can be more frequently found in the coastal areas, generally between the areas that experience low tide and high tide or they may be found in the sub-tidal region up to a depth where there is availability of 0.01% of light for the purpose of photosynthesis.

There a huge variety in the seaweeds and this can be attributed to the different kinds of plat pigments, exposure, light, temperature, depth of the seaweeds and the kinds of tides and their amount in addition to the characteristics of the shore. The classification of the seaweeds is majorly done on the basis of their color, which majorly go by:

1. Green algae (Chlorophyta);
2. Brown algae (Phaeophyta);
3. Red algae (Rhodophyta).

This classification of the seaweeds is different as it varies with the kinds of pigments they consist of and the kind of photosynthesis that they undergo, which later translates to the green, brown, and red algae.

What is evident is the fact that all the sea plants consist of chlorophyll, the pigment that helps in photosynthesis and in subsequent coloration of the plant. These result in categorizing the seaweeds into the three kinds which have been discussed hereafter.

1.1.1. Green Algae

The green algae are similar to the plants that are found on land. They contain chlorophyll in them and contain the two categories of it in ‘a’ and ‘b.’ The chlorophyll they contain is used to capture the light algae that they receive, for the purpose of photosynthesis. In turn, the chlorophyll pigments reflect green light as light falls on them, which makes their color appear to be green.

1.1.2. Brown Algae

These algae are mostly yellowish brown in color and have a greyish or a blackish tinge to them. This appearance of the brown algae can be attributed to them containing chlorophyll 'a' and 'c' in them. In addition to these kinds of chlorophylls, they contain an accessory pigment known as xanthophyll and are commonly called fucoxanthin.

This is a major contributor to these algae having a brown color in them. However, there are some sea plants or seaweeds that are completely black in their appearance. These algae are generally mistaken with the red algae, as they are mostly found to be dark in color.

1.1.3. Red Algae

In the red algae, there is the presence of only chlorophyll 'a,' which can be considered as the principle pigment for the purpose of photosynthesis. The red color of the red algae can be attributed to the series of accessory pigments which are commonly known as phycobilin (Figure 1.1).

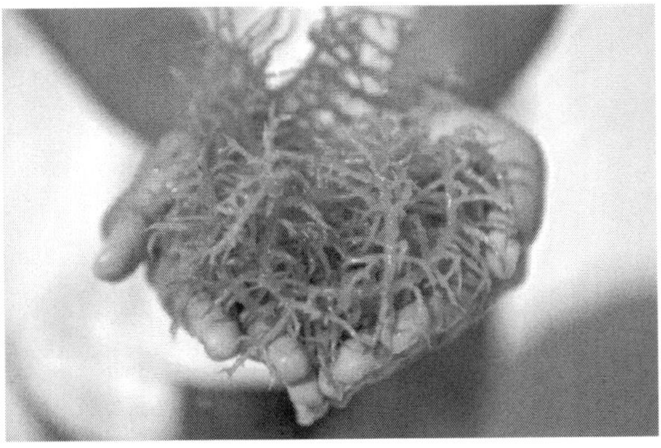

Figure 1.1. Seaweeds are good source of food products and rich in nutrients.

Source: Image from IAS Express.

These accessory pigments tend to absorb the blue and green light falling on them and reflect just the red light. However, even as they are known as the red algae, they may appear to be of various other colors as well such as yellow, black, blue, or even green.

This appearance of different colors of the red algae holds even more importance and is relatable in the case of intertidal red seaweeds. In cases pertaining to these, the red pigments of the seaweeds are masked by the chlorophyll and the other kinds of pigments that may be present in them.

The blue light has the maximum potential to penetrate the depth of the water as compared to any other light. Due to this, the phycobilin that are there in the red seaweeds allow the algae to carry out photosynthesis and hence make it possible for them to survive at the great depths that they are found at. And for a matter of fact, there is some seaweed that can be found even at the depth of 250 meters.

1.2. HISTORICAL OUTLINE OF THE ADVANCEMENT OF SEAWEED EXPLOITATION AND DEVELOPMENT

Human and seaweed collaborations appear to go back to the Neolithic period (Dillehay et al., 2008; Ainis et al., 2014; Erlandson et al., 2015), however the soonest written account of their human use starts from China, around 1700 years prior (Yang et al., 2017).

For a considerable length of time, beach front populations harvested a wide variety of seaweed from every single algal harvesting. At first, kelp was frequently utilized for household purposes as food and feed, while later, mechanical utilizations (gels, manures) (Delaney et al., 2016).

Early instances of usage of seaweed for restorative purposes incorporate the Chinese utilization of brown algae for goiter (sixteenth century, Chinese home grown, 'Pen Tsae Kan Mu'), Gelidium for intestinal tribulations and dried out Laminaria stipes for the enlargement of the cervix in troublesome childbirths (Levine, 2016).

The industrial utilization of seaweed biomass has moved throughout the years, from exploiting beach kelp as manures and a source of potash, by means of iodine production, to hydrocolloid extraction (Synytsya et al., 2015).

At all stages, the 'capability' of future industry has been seen as being bigger than its actual scale and this is as applicable today as it was 100 years before or more, when the business looked totally different (Hafting et al., 2015) (Figure 1.2).

Figure 1.2. Found at various depths of the seas, they need sunlight to variable extents.

Source: Image by New Food Magazine.

Numerous specialists predict a future where kelp will be developed for more important purposes than food items and feeds (Neori, 2016). These incorporate higher worth uses as raw materials for specific polysaccharides (for example agar, carrageenan, and alginates) (Bixler and Porse, 2011), or change of the biomass into items for specialized, strength agronomic applications (Buschmann et al., 2008; Craigie, 2011).

Moreover, further up in the worth pyramid one can discover useful items, for example, significant elements for food and feed (Fleurence, 2016), cosmeceuticals (Balboa et al., 2015), nutraceuticals (Himaya and Kim, 2015), pharmaceuticals (Thanh-Sang Vo et al., 2012; Cao et al., 2016; Anis et al., 2017), and bioalgae as a low worth yet exceptionally high volume application (Korzen et al., 2015; Fernand et al., 2016).

As these various applications can be reciprocal, diverse handling choices (Camus et al., 2016), including all-inclusive methodologies, have been developed.

The comprehensive 'biorefinery approach' consecutively removes the most significant parts from algal biomass, leaving the rest of the unadulterated material for product purposes, for example food, feed, manure, and fuel, while limiting waste and ecological effects of the procedure (Baghel et al., 2015; Trivedi et al., 2015).

1.2.1. Recent Production: Exploitation and Cultivating Procedures

China and Indonesia are by a long shot the biggest kelp producers with more than 23 million tons of collected production in 2014. China delivers mostly kelp for food (for example Saccharina japonica and Undaria pinnatifida), and red algae having a place with the genera Gracilaria and Pyropia (FAO, 2016). Then again, Indonesia develops mostly the carrageenophytes Kappaphycus and Eucheuma (FAO, 2016).

Taken together, these leading five genera-Saccharina, Undaria, Porphyra, Eucheuma/Kappaphycus and Gracilaria-represent approximately 98% of the world's developed seaweed production (Suo and Wang, 1992; Pereira and Yarish, 2008). Besides, Chile, China, and Norway lead the exploitation of the wild supplies of seaweed, of which kelps are the most looked for (FAO, 2016).

In 2014, the main seaweed cultivating nations, China, and Indonesia, each delivered in excess of 10 million tons, the Philippines and the Korean Republic more than 1 million tons, while the Popular Democratic Republic of Korea, Japan, Malaysia, and Zanzibar developed over 100 000 tons each.

In the Americas, just Chile has showed up in the cultivating measurements tables, with 12 836 huge amounts of developed Gracilaria. Generally, European, and African nations don't deliver seaweed except for a couple of nations that do develop minute amounts.

Interestingly, the European, Canadian, and Latin American kelp enterprises despite everything depend on harvesting common resources (Rebours et al., 2014).

In Chile, there is exploitation of the brown algae Lessonia and Macrocystis, and the carrageenophytes Sarcothalia crispata, Gigartina skottsbergii, and Chondracanthus chamisoii, yet cultivating is constrained distinctly to a generally limited quantity of Gracilaria chilensis (Buschmann et al., 2008).

In Portugal, reasonable production of kelp and seaweed-based items utilizes the integrated multi-trophic aquaculture (IMTA) idea, cultivating seaweed in vicinity to a few animal types at various trophic levels, permitting the decrease of aquaculture wastes (Troell et al., 2009).

IMTA is rehearsed in a controlled domain ashore, with natural confirmation for quality, recognizability, dependability of flexibly and a small carbon impression. Development of Ulva has been set up in South Africa, in light of a comparative IMTA idea (Bolton et al., 2009) and was received from a more minor perspective in Israel (Shpigel, 2013; Neori et al., 2017).

In Brazil, the business development of Kappaphycus alvarezii has been actualized during the most recent decade in coastal waters on the southern and south-eastern coasts (Pellizzari and Reis, 2011).

In Norway, seaweed exploitation (to a great extent Laminaria hyperborea and Ascophyllum nodosum) has a long convention (Meland and Rebours, 2012), and despite the fact that the nation was right on time to build up an information-based, incorporated seaside zone management framework, it was perceived that the developing interest of the business couldn't be fulfilled exclusively from the wild (Stévant et al., 2017).

To handle this issue the improvement of kelp aquaculture was started, in light of examination and pilot-scale production, and moved rapidly to get business grants for development of chiefly Saccharina latissima in coastal water (Stévant et al., 2017).

In the nations referenced above, wild seaweed exploitation has been completed economically for a considerable length of time (Rebours et al., 2014; Cottier-Cook et al., 2016).

However, over-reaping of Gelidium spp. in Japan (Fujita et al., 2006) and Morocco has enormously reduced Gelidium spp. beds, with conceivably destroying ramifications for the production of microbiology-grade agar (Callaway, 2015). Hence, endeavors towards development and substitution of wild harvest are much increasingly significant.

All in all, contingent upon their association (for example clonal versus unitary organisms) kelp are developed in one or different step cultivating frameworks (Santelices, 1999). Clonal species, for example, Gracilaria and Kappaphycus can be divided vegetatively and proliferated directly for development in culture frameworks (Figure 1.3).

Figure 1.3. The seaweeds are also helpful in containing the carbon in the oceans.

Source: Image by Oceana.

This should be possible at various scales, moving from intensive ashore tanks or lakes to broad vast sea culture frameworks utilizing long-lines or rafts. Then again, spread of the unitary seaweed, for example, kelp for industrial development requires an incubation center/nursery.

The distinctive development frameworks have been completely tried experimentally, however the most industrially fruitful frameworks have been those that culture kelp adrift, because of lower operational and capital expenses (Sahoo and Yarish, 2005).

1.3. GREEN ALGAE: DIVISION OF CHLOROPHYTA

The green algae are generally considered to be coming from the plants that are found on land for a common fact that they possess green pigments in the form of chlorophyll 'a' and 'b.' Some of the green plants that may be found in the best parts of the seas have been described hereafter (Figure 1.4).

Figure 1.4. Green algae on seashore rocks.

Source: Image by Science Photo Library.

1.3.1. Sea Lettuce (Ulva sp.)

The Ulva species are commonly known by the name of sea lettuce. These species are known to build dense groups of membranes which appear as blades and are as thin as just two cells combined. These green plants resemble lettuce leaves. The sea lettuce is used in various kinds of food items, mostly in the oriental cuisine.

1.3.2. Green Sea Intestines (Enteromorpha intestinalis)

These kinds of green plants appear to be made up of green tubes, which may be often found to be collapsed together and are as thin as just one plant cell. Their name is derived from the reference to tubes, hence the word 'intestines.'

Both-the sea lettuce and the green sea intestines-may be known to be plants that grow at a very rapid pace and change themselves according to the conditions and opportunities to grow and survive. They have a wide range of tolerance temperatures and also have great tolerance towards salt.

These features that they possess make them able to colonize successfully along the rocky seashores, if the conditions favor them. They can be found in high abundancy at high places on the shores where the main factor that controls the manner in which the sea plants will be distributed is the desiccation factor.

There may be a situation of sheltered conditions under which these kinds of sea plants act as the only sea plants that have the ability to occupy the high shore tide pools. In such cases, the sea lettuce plants may be able to grow to the lengths of more than half a meter.

These plants are the first plants in the sea to appear in scenarios where the various areas of the seashore may appear to be disturbed due to various factors as they have the ability to grow extremely fast and occupy the spaces that may be free for them.

1.3.3. Upright Codium (Codium fragile)

This seaweed is a kind of forked plant, mostly upright and has a velvety texture to itself. It is spongy and has a remarkable ability to control the manner in which its chloroplasts move from one place to another, so that it can maximize photosynthesis.

The thallus of this seaweed is generally made up of various filaments that are interlocked with each other and that become structures that are club-like in look and appearance. These filaments have the chloroplasts and the various reproductive parts in them.

Codium fragile is seaweed that may be grouped under the kind of sea plants that show uniqueness in the way that the filaments that are there in them do not have cross walls. Hence, at the time when they must be divided into cells, each of the filaments can be considered as a huge cell that consists of numerous nuclei. The Codium fragile may be abundantly found in the intertidal rockpools which show inappropriate proneness to the inundation of the sand.

1.3.4. Strap Caulerpa (Caulerpa filiformis)

This seaweed, Strap Caulerpa, is seen to grow as a group of different rhizomes that look similar to roots and are tangled together. This tangled mass results in bright grass that is flat and green in color and in the form of blades and bears cylindrical stipes.

The blade-like structures which are often found to be mottled with several spots, dark green in color, are further branched into two at one or two places. In a manner similar to that of the Codium, this seaweed also has an internal structure that is made up of filaments that contain more than one nucleus and does not have cross-walls, which thus goes on to be seen as lacking a cellular structure.

Even after having a non-cellular structure, the Strap Caulerpa shows good amount of toughness and exhibits a complicated morphology. This seaweed is able to retain the toughness and form because it is rich in cellulose ribs that are found in a complex way in it. These cellulose ribs fill up the space that it left due to the absence of cellular walls.

In addition to the toughness exhibited, it also reaps benefits from the absence of walls, in similar way as Codium, as the movement of the chloroplast is not restricted and it can control it to have maximum benefit from the photosynthesis process. Hence, these kinds of seaweeds are found in much abundance near the areas that are affected by sand inundation.

1.4. USES OF GREEN ALGAE

The sea plants are the basic step in the food web or the food hierarchy in the sea, along with the microscopic sea creatures-the phytoplankton. They also hold great significance for the humans. However, they do not have any smell or taste. Even then, there are many products that the humans use, which are derived from the sea and the seaweeds. Various kinds of uses of the green algae ha e been cited hereafter:

1. There has been a long time for which the Sea Lettuce has been a part of the cuisine in the far east of the globe and has been consumed in the form of green vegetables in salads and soups.
2. The algae are used as a source of the food colorant in food. This can be seen in the use of the Beta-carotene, which is a natural pigment that has been derived from the green algae. It is mostly used in:
    - i. Cheese;
    - ii. Egg substitutes;
    - iii. Mayonnaise;
    - iv. Margarine;
    - v. Coffee creamers;
    - vi. Multivitamins;
    - vii. Salad dressings and other such dishes.

In the current times, in South Africa, there is a good industry that is making use of the green algae. The manner in which the uses of sea plants have grown over the period of time, has compelled the state legislation to make laws for the sustainable use of these resources that can be of good commercial benefits.

This has also resulted in tempting the commercially backed companies to invest in more research on the seaweeds that are found locally, which has given positive results in the form of new and novel knowledge regarding the sea species.

1.5. BROWN ALGAE: DIVISION PHAEOPHYTA

The brown sea algae have been observed to be growing at a slower rate that the green algae. They have also been observed to have less tolerance towards the salt and the temperature ranges as compared to the green algae and are hence, more susceptible to the desiccation stress.

Thus, the brown algae have been found to be there on the lower side of the shore as compared to the green algae on the high shores. Some of the brown algae have been discussed and explained in the text hereafter (Figure 1.5).

Figure 1.5. Brown algae.

Source: Image from Seaweed.ie.

1.5.1. Cape Cord-Weed (Chordariopsis capensis)

The Cape Cord Weed is seaweed that has limpness and stringiness to itself and it found on the parts of the high shore that are mostly sheltered and in areas where there are mid-shore tide pools. In such areas, the weed becomes a mass that lacks tidiness and is flaccid in nature and has the tendency to be washed away by any kind of visible wave movements.

1.5.2. Dead Man's Fingers (Splachnidium rugosum)

This kind of seaweed contains branches that are full of spots and are cylindrical in shape. They are quite long and when looked closely, they appear as fingers that have withered away and have callused.

The branches appear to be berry-like sometimes and are always filled with viscous mucus that is clear in appearance. This mucus has a great significance for the plant as it enables the plant to withstand a good degree of desiccation stress which it may get exposed to at the times of low tides in the mid-intertidal regions.

1.5.3. Ralfsia (Ralfsia verrucosa)

Ralfsia can be found to be olive brown in color, which may later appear to be khaki. This kind of seaweed can be found in abundance in the pools that are found in the low shore tides. In such regions, the Ralfsia is able to grow to large expanses of land which are flat and smooth in appearance and feel. This tends to give an impression that someone has dropped large amount of khaki paint into the tide pools.

The Ralfsia grows at a really rapid pace and can be found in large abundance. It also specializes in the formation of a chemical that bars most of the grazers from not grazing it. However, there are some species that feed on the Ralfsia for their nourishment, such as the periwinkle and the long-spined limpet and they even have the tendency to look after their growth.

1.5.4. Hanging Wrack (Bifurcaria brassicaeformis)

This kind of seaweed is abundantly found in the parts of the low tide region that lie just above the subtidal zone. In these regions the hanging wrack is able to develop long and expansive mats of branches that appear tough and are cylindrical in shape and all this groups up to appear like a lawn that has not been mown.

This seaweed is able to dominate the seashore due to the fact that it spreads its holdfast in a rhizomatous manner. Some of the researchers and scientists claim that they have discovered sea plants that have diameters up to 1 meter.

This kind of species is very local to South Africa and is found in a limited way. It may be found only from the point of Cape Agulhas to the point of Sea Point, which is around Cape Town.

1.5.5. Long-Leafed Sargassum (Anthophycus longifolius)

The appearance of the long-leafed sargassum is that of knobble, robust, and twisted seaweed. It has the ability to easily dominate the large areas of the low shore tide pools and even the immediate areas of the sub-tides.

The kind of blades these species have may be taken to have a toothy characteristic as they have those blades near the surface and do not bear many bladders in the axils. These bladders help the seaweed to keep floating and help them to have maximum benefits in the photosynthesis process.

It is believed to be one of the largest non-kelp seaweeds, brown in color and can grow up to the lengths of 1 meter or even more at times. This plant is unique in the sense that it is the only species in the genus Anthophycus and is quite endemic to South Africa.

1.6. KELP

The kelp can be found abundantly in the subtidal and the intertidal areas of the sea shores. They are hugely grown plants and are brown in color. The kelps are similar to the trees in the old forests. The kelps can be found to dominate the subtidal zones near the shores, in the cool waters which are extremely rich in the nutrients.

The word 'kelp' was formerly used to refer to the ash that was formed from the sea plants. This ash was used in the form of fertilizers in Europe. The kelp can be seen as the fastest and the largest growing seaweed or plant and has the capability to grow as much as 13 millimeters in a single day.

There are some of the sea plants which are seen to grow more than 30 meters in length. These are the giant kelps that are found in abundance in the central and southern California and have the capability to grow by about 50 centimeters in a single day (Figure 1.6).

Figure 1.6. Kelp forests.

Source: Image by Sport Diver.

The kelps differ from the other sea plants in a way that they are not as simple in structure as the other sea plants. The kelps grow in a similar fashion to the higher green plants. Due to their structure being so large in size, they tend to develop good tissues and organs for themselves.

The tissues that have been developed in such a specialized manner help in transportation of the necessary things for the purpose of photosynthesis in the internal movements. However, there are other tissues such as photosynthetic, strengthening, and reproductive tissues.

The specialized organs in the kelps mostly involve:

A holdfast that has a similar appearance as roots and is solely for the attachment.

A stipe similar to a stem in a plant which contains the blades pertaining to photosynthesis and reproduction.

Floats that help the sea plants and the sea plants and the blades remain erect and afloat in the water column.

All the features that have been mentioned above have been very instrumental in improving the photosynthetic ability of the kelps, which has enabled them to develop good lengths of beds of themselves in the subtidal regions and remain as a dominant plant in those regions. Some of the popular kelp species have been discussed hereafter.

1.6.1. Sea Bamboo (Ecklonia maxima)

Ecklonia maxima, commonly known as the sea bamboo, is a species that is found in most abundance and is seen to dominate the inshore regions in many parts of the world. This seaweed is seen to form a huge holdfast which may extend up to 15 meters in length and in the form of a hollow gas filled stipe. This stipe may end up in the form of a bulb at its end.

The bulb-like structure then goes on to extend in the form of a flat and solid primary blade and from these blades emerge the secondary blades. The secondary blades emerge up to a length of 3 meters. As this species contains a hollow stipe and a bulb, it has buoyancy acting on it which keeps them afloat in the ocean.

The blades of this species keep coming up on the surface of the oceans and can be seen so in the low tide. Thus, if in case anyone sees kelps on the surface of the ocean, they may presume it to be sea bamboo in most cases (Figure 1.7).

Figure 1.7. Sea bamboos.

Source: Image by iSpot.

This thing also suggests that the sea bamboo forms the basis for the canopy of the forest formed by the kelps and it is in this canopy that one can find three species of the red sea plant along with a limpet that may have grown on the sea bamboo epiphytically.

This species may be regarded as a dominant species in the areas constituting the inshore waters. However, in subsequent times, it gets replaced by the other kelp species in the deeper waters known as the split-fan kelp.

1.6.2. Split-Fan Kelp (Laminaria pallida)

Laminaria pallida does not have the buoyant forms found in sea bamboo (in spite of the fact that the stipes of a few plants have been demonstrated to be hollow), so it once in a while breaks the surface in deeper water.

This species develops to around 10 m long and has a solitary expansive, fan-molded blade that turns out to be sporadically split, giving the kelp its name. In contrast to the sea bamboo, which is increasingly regular in generally shallow water, split-fan kelp happens to be found at depths of 30 m.

Grown-up split-fan kelp plants have warty stipes while they are smooth in the sea bamboo. Both of these sea plants are regular on rough shores exposed to waves.

1.6.3. Bladder Kelp (Macrocystis angustifolia)

The bladder kelp, (Macrocystis angustifolia) is minimal basic of the west coast kelp species. This kelp is a moderately sensitive, vine-like species bearing various sharp blades at regular stretches along its thin stipe.

At its peak, the new blades are melded, slowly isolating as they develop. The sharp blades have various negligible spines and a solitary gas-filled bladder at their bases. It is these bladders that permit this kelp to stay above water and erect in the water segment.

In contrast to the next two types of kelp, an individual rhizomatous spreading holdfast in this species can tolerate different stipes. Huge populations of this sea plant are discovered uniquely at two areas close to Cape Town; one at Kommetjie, the other at Robben Island in the Melkbosstrand district.

Distinct to this type of kelp are its heedlessly undulated sharp edges. This element is particularly significant, as it takes into consideration expanded ingestion of nutrients in generally low stream situations average of shielded tidal ponds or embayment.

1.7. USES OF BROWN ALGAE

A trademark highlight of brown sea plants is that their cell walls are made of cellulose fortified by calcium alginate. Alginate is a substance of significant financial importance as it is utilized as a gelling and emulsifying operator in various industries.

However, at present, just kelp is reaped industrially in South Africa for alginate extraction. Maybe it is on the grounds that no other group of brown sea plant happens in large enough densities to be monetarily suitable. In South Africa, the sea plant industry depends on sea bamboo and split-fan kelp, however incorporates some other brown and red sea plants.

1.8. USES OF KELP

Globally, an enormous market exists for the reaping of kelp and other brown sea plants for alginate production.

1.8.1. Compost

Kelp is generally utilized as compost and is harvested broadly as feed for monetarily cultivated abalone.

1.8.2. Nutrition Enhancement

Sea bamboo is likewise utilized as a wholesome nutrient for livestock. In local farming, this kelp is notable for the creation of a fruitful plant development energizer. It is likewise an extraordinary source of micronutrients.

1.8.3. Food Industry

It makes water-based items thicker, creamier, and increasingly stable over extreme contrasts in temperature, pH, and time. For instance, alginates forestall the formation of ice precious stones in dessert. Different items containing alginate incorporate:

- Brownie mix;
- Frozen foods;
- Desserts;
- Relishes;
- Serving of mixed greens dressing (salads);

- Sauces;
- Gravies;
- Even lager.

1.8.4. Cosmetics, Pharmaceutical, Paint, Material, and Welding Industries

Alginate helps in the suspension and balancing out of operators over scopes of temperature and pH. In the paper business, for instance, alginate is significant as it empowers measuring and cleaning of the finished paper item. Did you realize that alginate from kelp is even used to make filaments for great sound speakers?

1.8.5. Medical Industry

Kelp is particularly significant as its alginate is utilized to epitomize numerous things, for example, tablets in powder form, crack castings and forms, and even organs for transplant. Kelp likewise contains a surprising measure of nutrients and minerals.

The most significant of these enhancements however is presumably iodine. Iodine has a normalizing impact on the thyroid organ (this organ controls the body's development and improvement). In Namibia, researchers are hoping to utilize sea bamboo and split-fan kelp as a treatment of goiter and avoidance of (or decrease in the event of) cretinism in southern Africa.

Correspondingly, in light of the fact that iodine takes care of the thyroid, kelp is even utilized in weight reduction equations. Japanese examinations have even indicated an immediate connection between alginate contained in kelp and the prevention of breast cancer.

Naturally, the kelp forest gives a significant, intricate, three-dimensional environment for various types of fish, invertebrates, and other sea plants. The kelp limpet, Cymbula compressa for instance, is found just on sea bamboo. Kelp beds are among the most profitable environments on earth, supporting high essential production levels.

This high efficiency frames the base of numerous seaside food networks in cool water situations around the world. Along these lines, both straightforwardly and in a roundabout way, they are a significant food hotspot for an enormous variety of invertebrates, fish, vertebrates, and seabirds.

Besides, kelp is intense and versatile, and extending into the sea, generally for some kilometers, they help break the incredible power of the waves offering insurance to the close shore biological system.

1.9. KELP FOREST FORM (FIGURE 1.8)

Figure 1.8. Kelp forest forms.

Source: Image from Google Sites.

1.9.1. Canopy Species

In the kelp beds found on the southern-west coast, the blades form a canopy layer close to the surface. The canopy retains a significant part of the light arriving at the water surface, so the vast majority of the stipes of develops plants is in a strange place.

1.9.2. Epiphytes

Like creepers and ferns found on the land, marine epiphytes grow in a way joined to the kelp blades and segments of the stipes close to the surface thus getting as much light as the edges. Epiphytes are commonly sensitive in nature, as they need to twist as the kelp influences in the waves.

1.9.3. Under Storey Species

On the stone, underneath the kelp canopy happens various types of sea plants that make up the understorey layer. These species live in a dim world, since the canopy and the epiphytes catch a significant part of the approaching light.

In any case, there is sufficient light to help a wide variety of understorey vegetation, a significant number of which can photosynthesize even at exceptionally low degrees of light. A significant number of these understorey sea plants are flat in shape and structure, to catch however much light as could reasonably be possible for them.

1.10. FOOD WEBS IN THE KELP FOREST

The South African kelp environment has been depicted as a sun-fueled framework, with extra energy contribution to the type of wind and waves:

Energy from sunlight drives essential creation, in addition to setting up a thermally delineated water section.

Wind energy drives the upwelling procedure, which brings nutrients up to the surface waters where the kelps develop. The take-up of nutrients is improved by water movement because of wave activity, and this permits the kelp and related sea plant to utilize the sunlight for essential creation.

Energy from waves can likewise remove kelp plants and wash them up onto the shore, where they give food to a wide variety of life forms, particularly amphipod and isopod crustaceans. Wave energy adds to the disintegration of the blades, delivering particulate and dissolved organic matter.

Wave activity keeps kelp plants moving, and keeps nibblers from climbing and harming the plants.

1.10.1. Energy Flow

This is the development of energy through the food web of a biological system, from sunlight to the essential producers, and from essential producers to consumers and decomposers. Kelp energy isn't utilized distinctly inside the kelp bed; it additionally assists with supporting other contiguous biological systems.

Each winter the South African west coast is covered by huge waves emerging from south Atlantic storms. A lot of kelp is removed, and cleaned up onto the shore, where it decomposes and is eaten by amphipods, isopods, and fly larvae.

1.11. RED ALGAE (DIVISION RHODOPHYTA)

The mid to lower intertidal of the Cape Peninsula is ruled to a huge degree by red algae. These form by a wide margin the biggest group of sea plants on the Peninsula.

Like the brown algae, a large number of the reds are beefy and massive and for the most part happen let down the shore since they are less tolerant of saltiness and temperature limits when contrasted and a portion of the normal green sea plants. The following are the common red algae of the Cape Peninsula (Figure 1.9).

Figure 1.9. Red algae.

Source: Image by Hindustan Times.

1.11.1. Purple Laver (Porphyra sp.)

Porphyra species, generally called purple laver, are very thin (just a single cell thick), level, membranous plants. This sea plant seems as though a sheet of wrinkled cellophane when dry, folded, and collapsed upon itself.

They differ in color from yellow to purple to practically dark. Different types of Porphyra are in abundancy high up on the shore where drying up stress is at its most prominent.

While most sea plants are prejudiced of the conditions this high up the shore, Porphyra species can make due here because of their astounding recuperation capacities; this sea plant has been appeared to endure dehydration for more than seven days. Purple laver is frequently harvested to use as the folding over sushi and can be dried, ground up and improved to make nori sheets, which are also utilized for sushi.

1.11.2. Tar Crust (Hildenbrandia lecanellierii)

Another species lenient of the parching stress high up on the shore, is the encrusting red sea plant (Hildenbrandia lecanellierii). Looking amazingly like sprinkled tar, this sea plant may shape huge scopes in the upper intertidal, in crevices, and in places prone to sand immersion.

More established pieces of the plant regularly become segregated from the substrate, yet it shows an astounding pace of recuperation from physical aggravation and frequently may regenerate from a couple of cells.

1.11.3. Slippery Orbits (Aeodes orbitosa)

Ordinarily known as Slipper orbits this is another level, tricky, foliose sea plant that is additionally intense. Its color differs from yellow brown to reddish brown. This sea plant is particularly basic in the mid-to lower intertidal zone where it frequently forms huge regions. In protected intertidal pools, this sea plant has been known to develop to 2 meters in distance across.

1.11.4. Tongue Weed (Gigartina polycarpa)

This is an intense, plump sea plant bearing oval edges. People of this species are distinctively harsh and bear various papillae (some of the time, the sharp blades are even undulated) giving the presence of an unpleasant tongue. The color in this species changes from yellow-brown to ruddy brown to practically black. Alongside Aeodes it is likewise very copious in the mid- to lower intertidal zone.

1.11.5. Hedgehog Sea Plant (Nothogenia erinacea)

Suitably called the hedgehog sea plant due to its appearance, this sea plant comprises of lengthened foliaceous sharp blades that are extreme and rough when wet, and get papery when dry. The sharp blades bear various densely packed tuft like outgrowths giving the sea plant its hedgehog-like appearance. This species changes in color from yellow-brown to practically dark. It is additionally extremely normal in the mid to lower intertidal.

1.12. USES OF RED ALGAE (FLESHY)

While the algae industry in the Western world depends for the most part on sea plant removes, in the East Sea plants are developed in gigantic volumes for human utilization. Quite a bit of this industry depends on the red sea plants. Of all the sea plants, the reds are likely the most significant to people financially.

Purple laver (Porphyra) comprises an amazing 80% of all sea plant harvesting everywhere throughout the world. In Asia, purple laver is referred to as nori and is eaten as complete plant either dried or in soups, or as scrumptious wrappings for sushi and rice.

The iodine and high nutrient and protein substance of norimakes it alluring, as does the general straightforwardness of its Mariculture (sea cultivating), which started over 300 years prior in Japan. Other than food for direct utilization, red sea plants are likewise significant for their phycocolloid extracts.

Phycocolloids are sea plant subordinates that cause particles to stay suspended in solution and are in this way fantastic as stabilizing and gelling specialists. The primary phycocolloids got from red sea plants are carrageenan and agar.

Carrageenan is exceptionally looked for after in western societies where it is particularly significant in the dairy business. Milkshakes, cheese, yogurts, powdered milk (counting child recipe), and so on, all have red sea plant separates.

In fact, carrageenan is even utilized in pet food, beauty care products, shampoos, paints, and toothpaste. Various carrageenophyte sea plants are being examined in South Africa. These incorporate green tips and different types of tongue weed, turned gigartina, and elusive circles.

Agar then again, has its most significant use as a medium on which to culture fungi and bacteria in microbiological and Medical pathology research. In nourishment for human utilization agar can be found in processing and confectionary items and is likewise broadly used to explain wine, juice, and vinegars due its phenomenal protein restricting properties.

In bigger businesses, agar is utilized to make adhesives and capsules for tablets. South African agarophytes incorporate the sea plant genera Gracilaria (which is harvested broadly in Saldanha Bay) and particularly Gelidium.

1.13. SEAWEED CULTIVATION AND MODELING PRODUCTIVITY

Until this point in time, kelp just produce a small part of the worldwide supply of biomass with underneath 30 $\times 10^6$ fresh weight (FW) tons of seaweed, in contrast with 16 $\times 10^{11}$ huge amounts of yields from the land, grasses, and forests.

An extending body of proof proposes that off-shore developed seaweed, which don't contend with food crops for arable land or consumable water, could give an alternative source of biomass for the practical production of food, synthetics, and powers (Radulovich et al., 2015; Lehahn et al., 2016; Neori, 2016).

In any case, increasing the worldwide production of seaweed requires a comprehension of the critical points that right now limit their production.

A few investigations developed models for evaluating the net primary productivity (NPP, characterized as g DW m^{-2} day^{-1}, an estimation of the proficiency of transformation of sunlight based to conceivably helpful chemical energy) of chosen kelp for biological and ecological applications.

Duarte and Ferreira (1997) distributed a powerful model, which anticipated NPP for Gelidium sesquipedale. The physiological model was helpful for the administration of common loads of this species as raw material for different businesses.

The model thought about profundity, profitability as a component of light energy, temperature, and rate of respiration, commodity exudation, frond breakage, and mortality and incorporated a transformation factor among carbon and dry weight (DW).

Another model was developed for the streamlining of harvesting Ascophyllum nodosum from characteristic populations (Seip, 1980) that took into consideration the assurance of the ideal thickness of seaweed inside the population, and the rate and recurrence of collecting, permitting just 7% of the evaluated standing yield to be reaped, to deliver a manageable harvest over a time of quite a few years of exploitation (Lee and Ang, 1991; Ugarte and Sharp, 2001).

A linear model for lake cultured Gracilaria conferta indicated a positive connection among development and temperature, where development took place distinctively in various seasons, and where nitrogen fixation in the surrounding seawater restrained advancement of epiphytes, under states of week after week nitrogen beat taking care of (Friedlander et al., 1990; Friedlander, 1991).

These models and their responses to harvesting are of crucial significance for the improvement of strategies required for those pieces of the business seaweed exploitation industry which are as yet dependent on characteristic resources.

Notwithstanding how supportive these models may be, they can't be utilized legitimately for demonstrating kelp farming destinations yet, as ecological conditions in those development locales are not the same as those happening in nature (as utilized by the model).

Estimations for the worldwide capability of off-shore seaweed farming are uncommon. Notwithstanding, a numerical model for Ulva spp. digestion and development rate was as recently developed (Lehahn et al., 2016).

The model ran on a worldwide 1° matrix with one yield record for every long period of the year. Algal development rate (μ) was determined as an element of:

- Light force (I);
- Temperature (T);
- Salinity (S);
- Dissolved nutrients (nitrate and phosphate);
- Respiration rate.

The model accepted that every one of the natural elements differently affected the development pace of the biomass. This model, in light of oceanographic information, was utilized to distinguish the possible worldwide profitability of Ulva spp. in locales up to 400 km from the shore, related with up to 100 m deep securing establishments and characterized as 'near-future, deployable biorefinery provinces' (NDBP) (Lehahn et al., 2016).

The model gave an estimation of the hypothetical, potential biomass production as 10^{11} DW tons per year, over a surface territory of ~10^8 km^2. By utilizing one of the most beneficial seaweed, for example, Macrocystis pyrifera, it was demonstrated that an aggregate of 200 wet tons per ha per year could be acquired (Buschmann et al., 2014; Correa et al., 2016).

Notwithstanding, when scaling up, production effectiveness could drop with huge variations in existence (Camus et al., 2016). These researches call for extraordinary alert to be practiced as certain outcomes may be somewhat unacceptable and what was seen as 'high potential' may not turn into a 'useful reality.'

Lehahn et al. (2016) recommended that a significant constraining component for enormous scope seaweed production was biomass yield.

Along these lines, under ordinary sunlight conditions, for example, outdoor development, the quantity of photons doesn't restrain photosynthetic rate, rather (as demonstrated by various other seaweed examines) the pace of photosynthesis was restricted by different physiological procedures, for example dissemination and inherent plant and carbon-obsession digestion systems (Hurd et al., 2014).

Studies demonstrated that equivalent photosynthetic rates could be accomplished by consistent and beat light with explicit frequencies (Tennessen et al., 1995). This property has been generally utilized in the form of on-shore, algal photobioreactors, where mixing is controlled to strengthen the complete miniaturized scale and macroalgal yields (Bidwell et al., 1985; Buschmann et al., 1994; Bruhn et al., 2011).

Mixing has, as a rule, been expected to improve nutrient dispersion for the algae and to streamline light presentation (Msuya and Neori, 2008). Another idea for off-shore development of seaweed builds yield per unit territory by applying outer mixing to exploitation normal photon catch and carbon obsession rates.

A definitive objective of this framework was to build the total energy effectiveness of off-shore kelp production by streamlining of photon usage per unit zone for an off-shore homestead (Golberg and Liberzon, 2015).

Mixing would use the enormous volume of the coasting off-shore kelp reactor by presenting algal cells to solar energy for a brief timeframe to catch photons at that point moving the algae to a characterized depth for a while for carbon fixation so another layer of seaweed would be presented to accessible sun, depicting what occurs in high rate production lakes ashore. It is a practical approach that can accomplish the desired results and is being investigated and developed (A. Golberg, individual correspondence).

Notwithstanding the innovative enhancements required for huge scope development, huge extra endeavors and resources are required to develop and choose seaweed species and strains with explicit properties customized for food, compound, or fuel applications (Zhang et al., 2007; Robinson et al., 2013; Li et al., 2016).

Mechanical strain advancement in agrarian yields was changed by the presentation of high-throughput phenotyping with imaging. Imaging spectroscopy and thermal infrared-, fluorescence-, 3D- and tomographic imaging have been utilized to effectively decide the necessary properties of terrestrial plants for strong, financially savvy, downstream processing (Cabrera-Bosquet et al., 2012; Li et al., 2014; Mutka and Bart, 2015; Walter et al., 2015) and have additionally been utilized for algae (Shefer et al., 2017).

In any case, the adequacy of these and other current instruments in increasing seaweed efficiency on a business huge scope adrift has still to be illustrated.

1.14. SOCIAL AND ECONOMIC EFFECTS AND CHALLENGES

Harvests from wild populations are influenced by overexploitation and climatic changes. In Northern Ireland, for instance, which is recorded as one of the best 10 producers of wild stocks all around (FAO, 2014), McLaughlin et al. (2006) portrayed in detail the unfriendly effects of seaweed harvesting at small, distinctive, and business scales on regions of preservation significance, protected, and priority habitats and species, including:

- Unsettling influence of birds and wildlife;
- Disturbance of food webs;
- Harm to substrata;
- Habitat destruction;
- Confined biodiversity changes;
- Changes in particle size appropriation in residue.

Direct consequences for the kelp population incorporate mortalities because of increased development rate and cover of other algae which are not harvested, for example, filamentous algae, and the brown seaweed, Fucus vesiculosus, which outcompete the ideal species, and die-back because of increased predation.

In a few regions of Norway, the kelp Saccharina lattisima has been accounted for by Moy and Christie (2012) to have endured dieback by 40–80% because of sea urchin predation. The brown seaweed kelps are generally influenced by rising water temperature, on the grounds that sexual reproduction (gamete development) in many types of kelp won't happen above 20°C (Dayton, 1985; Dayton et al., 1999).

As of now along the European coasts and particularly in Brittany, France, the brown kelp, Laminaria digitata, which is vigorously reaped for business uses, is accounted for to be very close to local extinction.

The effectively decreased conceptive capability of the kelp because of diminishing population and harvesting instigated biological system changes might be exacerbated by climate caused increment in sea temperature (Brodie et al., 2014; Raybaud et al., 2013).

Two other kelp species, Laminaria ochroleuca, a warm-calm enduring, and Saccorhiza polyschides, a wide-going cool-to warm-temperature yearly, have, to some degree, higher temperature resistances for sexual reproduction than different kelps (Pereira et al., 2011); however, Saccorhiza outcompetes L. ochroleuca in shared natural habitats.

Brittany is the northern furthest reaches of L. ochroleuca's range. Since 1940, L. ochroleuca has been found on the shores of southern England, which is obviously demonstrative of a moderate northward expansion of hotter waters.

Foreseen increasing sea temperatures later on in the boreal area may bring about L. ochroleuca potentially supplanting L. hyperborea (Brodie et al., 2014). Then again, the kelp Ecklonia maxima are increasing eastbound on the tip of South Africa in view of a northward interruption by cooler inshore water (Bolton et al., 2012); this could enormously profit the entire biological system and give more food to the abalone business there.

This is a serious complexity from southward interruption designs by warm water on the east and west banks of Australia, causing broad retreat of kelps and fucoid (another harvesting of brown algae) southward from their past northern-most cutoff points (Wernberg et al., 2011; Millar, 2007).

Seaweed farming and culture are truly influenced by infections. Ice-ice disease has affected the farming of the kappa-carrageenan-delivering Kappaphycus alvarezii, industrially called "cottony." Another species, Eucheuma denticulatum, industrially called "spinosum," is ice-ice-safe, however contains particle carrageenan which brings a much lower cost on the world market (Valderrama, 2012).

This issue might be a consequence of the low hereditary variety in K. alvarezii, the entirety of whose cultured stocks around the world have a comparative mitochondrial haplotype, which isn't the situation for E. denticulatum (Halling et al., 2013; Zuccarello et al., 2006).

Noteworthy sicknesses influencing developed kelps (e.g., Saccharina japonica) incorporate green-decay, white-decay, rankle infection, which might be ecologically initiated, and contortion ailment of summer sporelings and swollen stipe or "frond curve ailment" which are brought about by microscopic organisms (Brinkhaus et al., 1987; Tseng, 1986).

Parasites, for example, Pythium, an oomycete growth, causes "red rot" or "red wasting" ailment in the red kelp Pyropia, generally utilized in making sushi (Hurd et al., 2014). However, in view of contextual studies from six nations, Valderrama (2012) detailed that the financial effects of seaweed farming have been sure.

He credited this essentially to small scope, family tasks, bringing about the creation of generous jobs when contrasted with different types of aquaculture.

He included that seaweed farming is frequently attempted in remote territories where waterfront networks face less financial other options and where a large number of these networks have customarily depended on beach front fisheries which are at present being influenced by overexploitation.

Valderrama expressed that the effect of seaweed farming in these cases goes past its prompt economic advantages to communities as it additionally lessens the impetuses for overfishing. However, one test looked by farmers in these remote territories is low benefits because of high delivery costs.

This drawback is exacerbated by the reliance of farmers on processors for the acquirement of their farming materials and their absence of homestead the board abilities. Moreover, food safety issues can some of the time influence markets and costs. This is on the grounds that seaweed is effective nutrient extractors (Kim et al., 2014) and may aggregate compounds that can cause damage to human wellbeing.

1.15. CONCLUSION

From the chapter, it can be understood that there are three kinds of seaweeds- Green, Red, and Brown and all three of them have their own industrial or other kinds of uses. The growth and abundance of the different kinds of seaweeds is dependent on the amount of sunlight they need for the process of photosynthesis and the amount of water they need to grow and remain hydrated.

Some of the important species of seaweeds have been discussed in detail to elaborate on the requirements and the diversity in the abundance of seaweeds.

REFERENCES

1. (2016). Chapter 14: Seaweeds. [eBook] United Nations. Available at: https://www.un.org/Depts/los/global_reporting/WOA_RPROC/Chapter_14.pdf (accessed on 10 July 2020).
2. (n.d.). Module 3: Sea Plants. [eBook] Available at: https://www.aquarium.co.za/uploads/files/Module_3_-Sea_Plants.pdf (accessed on 10 July 2020).
3. Buschmann, A., Camus, C., Infante, J., Neori, A., Israel, Á., Hernández-González, M., Pereda, S., Gomez-Pinchetti, J., Golberg, A., Tadmor-Shalev, N., & Critchley, A., (2017). Seaweed production: Overview of the global state of exploitation, farming, and emerging research activity. European Journal of Phycology, [online] 52(4), 391–406. Available at: https://www.tandfonline.com/doi/full/10.1080/09670262.2017.1365175 (accessed on 10 July 2020).
4. Marine-aquaculture.extension.org. (2019). Seaweed Aquaculture: An Introduction-Marine Aquaculture. [online] Available at: https://marine-aquaculture.extension.org/seaweed-aquaculture-an-introduction/ (accessed on 10 July 2020).
5. Niobioinformatics.in. (n.d.). Introduction to Seaweeds. [online] Available at: http://www.niobioinformatics.in/seaweed/introduction.htm#:~:text=Introduction%20to%20seaweeds&text=Seaweed%20or%20benthic%20marine%20algae,they%20photosynthesize%20and%20produce%20food (accessed on 10 July 2020).

CHAPTER 2

Sources of Seaweeds

CONTENTS

Consumption of seaweeds has been increased over the years; hence, its demand has risen over time. It is mostly consumed in Asian countries. It is a rich source of nutrients for marine life as well. Most of the marine animals are dependent on these seaweeds. Here, in this chapter, the sources of seaweeds have been evaluated. It describes different types of seaweeds and their sources of production. This chapter also evaluates the consumption patterns and trends in the consumption of seaweeds and its industrial application.

2.1. INTRODUCTION

Seaweed may have a place with one of a few groups of multicellular algae: the red algae, brown algae, and green algae. As these three groups are not thought to have an ancestor which is typically multicellular, the seaweeds are a polyphyletic group.

Additionally, some tuft-shaping blue-green algae (Cyanobacteria) are now and then considered as seaweeds. "Seaweed" is a casual term and does not have a conventional definition. Two explicit natural necessities overwhelm the seaweed environment.

These are the proximity of seawater and the closeness of light adequate to drive photosynthesis. Another basic necessity is a firm connection point. Thus, seaweeds most ordinarily occupy the coastal zone and inside that zone, more every now and again on rough shores than on sand or shingle.

Seaweeds possess a wide scope of biological specialties. The most noteworthy height is just wetted by the highest points of ocean shower; the least is a few meters down. In certain territories, littoral seaweeds can stretch out a few miles out to the ocean.

The constraining component in such cases is daylight accessibility. The most profound living seaweeds are a few types of red algae. Various species, for example, Sargassum have adjusted to a completely planktonic specialty and are free-skimming, contingent upon gas-filled sacs to keep up an adequate profundity.

Others have adjusted to live in flowing stone pools. In this living space, seaweeds must withstand quickly changing temperature and saltiness and even incidental drying.

In 1999, WLW distributed a rundown of world seaweed use (Zemke and Ohno, 1999). At that point, dependable information on seaweed reaping and cultivating were hard to get a hold of, thus the information was separated from the book Seaweeds of the World, altered by Alan Critchley and Masao Ohno.

This asset was refreshed in 2006 and delivered as a CD-ROM entitled World Seaweed Resources (Critchley et al., 2006). Since this time, the Food and Agriculture Organization (FAO) of the United Nations has incorporated and made openly accessible a dataset of every detailed fishery and aquaculture arrivals from 1952 to the present, the most recent information is as of now from 2012.

While almost certainly, this information is to some degree under-announcing the arrivals of different species, it is the most present and precise dataset accessible. To give a report on Zemke-White and Ohno (1999), we got the detailed seaweed creation loads from the FAO (2014) dataset.

The detail of the announced seaweed characterization went from phylum level directly down to singular species. At the point when the seaweed was not answered to the species level, we utilized Critchley et al. (2006) to clarify genera or species where conceivable.

The writing utilized for every nation is provided here. An immediate examination between Zemke-White and Ohno (1999) and this present information is troublesome as the previous announced dry weight (DW) arrivals, while the FAO report utilized wet weight.

However, utilizing a dry: wet weight transformation, a few correlations are conceivable. Unquestionably, the genuine species that are used and their source nations can be thought about for changes in the course of the last 10 to 15 years.

As referenced before, the manner in which information are accounted for in the FAO dataset implies that it isn't generally conceivable to decide the species being accounted for. Once in a while just the more significant level grouping of Chlorophyte, Phaeophyte, or Rhodophyte (for green, brown colored, and red seaweeds, individually) was accounted for.

There are additionally issues with specific species groupings. For instance, different types of the genera Eucheuma and Kappaphycus are cultivated widely and practically just for the creation of carrageenan.

2.2. TYPES OF SEAWEEDS

Seaweeds are plentiful in nutrients, minerals, dietary strands, proteins, polysaccharides, and different practical polyphenols. Numerous scientists have concentrated on brown algae as a potential wellspring of bioactive materials in the previous scarcely any decades.

Seaweeds has been recognized as a potential maker of wide range of characteristic substances, for example, carotenoids, fucoidans, and phlorotannins demonstrating distinctive natural exercises in fundamental mechanical applications including pharmaceutical, nutraceutical, cosmeceutical, and useful food. This survey centers around natural exercises of the brown seaweed seaweeds dependent on most recent research results, including cancer prevention agent, anticoagulative, antimicrobial, antihuman immunodeficiency infection, mitigating, immunomodulatory, antimutagenic, antitumor, and anticancer impacts. The realities summed up here may give novel experiences into the elements of seaweeds and its subordinates and possibly empower their utilization as practical fixings in potential mechanical applications.

2.2.1. Red Seaweeds

Red seaweeds have had a more different advancement than the green and the brown. Numerous species can't stand drying up and overwhelm the between flowing stone pools. Others endure drying up, for example, the purple laver which can regularly be seen loosened up like a dry dark film over mussel beds on rough sea shores. Red seaweeds, for example, Polysiphonia lanosa are epiphytes; these are plants that develop on different plants for physical help. For this situation the epiphyte profits by the host's lightness lifting it closer to the daylight (Figure 2.1).

Figure 2.1. Red seaweed.

Source: Image by Wikimedia.

The red shade of the seaweeds is because of the bigger measure of red phycobilin colors abrogating the green color chlorophyll. The principle biomass of red algae overall is given by the Corallinaceae and Gigartinaceae.

The red algae Gelidium, Gracilaria, Pterocladis, and other numerous red algae are utilized in the production of the significant agar, utilized generally as a development vehicle for microorganism and other biotechnological and food applications.

Another significant red seaweed alga is Eucheuma utilized in the creation of carrageenan, a significant item utilized in beautifiers, food handling, and modern uses, just as a food source.

Probably the most noteworthy caargeenan species incorporate Betaphycus gelatinae, Eucheuma denticulatum, and a few types of the sort Kappaphycus including K. alvarezii (Lobban and Harrison, 1994).

The third most elevated delivered seaweed as far as biomass is Porphyra spp., having a complete creation of 1.8 million tons. China is the biggest maker of this seaweed, trailed by Japan and South Korea (FAO, 2014).

A few endeavors to build up Porphyra cultivating in the United States and the United Kingdom are directly going on, yet their commitments are seen to be lesser when contrasted with Asian nations (Chen and Xu, 2005).

Chondrus crispus is a red seaweed run of the mill to calm waters and has been a mainstream wellspring of carrageenan since the mid-1970s. The seaweed is generally collected from the wild and is to a great extent delivered in Canada.

It is viewed as the most seasoned seaweed industry in the United States. The most recent creation information from the FAO was in 1999, when an aggregate of 81 tons of C. crispus were delivered.

At present, C. crispus is refined industrially by Acadian Sea plants, LLC in Canada (Chopin and Ugarte, 2006). What's more, some presented types of red seaweed have become so obtrusive that their expansion has brought about high biomass creation.

This incorporates Asparagopsis armata in the North-East Atlantic and Mediterranean, Gracilaria salicornia in Hawaii, and Kappaphycus spp. in Hawaii, Brazil, and Fiji (Trowbridge, 2006; Zemke-White and Smith, 2006).

2.2.2. Brown Seaweeds

Laminaria sp. 'kombu', Undaria sp. 'wakame', Hizikia fusiforme 'hiziki' is consumable and a significant asset in Asian nations such as especially China and Japan. They are expended crude, bubbled, or dried material with improved green beans, jam, squashed ice, and coconut milk in Southern Vietnam (Tsutsui et. al., 2005).

Laminaria sp. was in abundant gracefully in Japan, predominantly from the northern island of Hokkaido, where a few normally developing animal categories were accessible. Undaria sp. has been gathered from normal assets for a long time in China, Japan, and Korean locale (Figure 2.2).

Figure 2.2. Brown seaweed.

Source: Image by Wikimedia.

Another algae Cladosiphon okamuranus 'mozuku' as plate of mixed greens in Okinawa-Japan (Thoma, 1997; Zhang et al., 2007; Zhu et al., 2009). Sargassum sp. is known as horsetail and it is eaten as soup or dressed with soybean sauce, or subsequent to being handled in Korea (Madlener, 1997) and in Hawaii (Novaczek, 2001).

In the Pacific area, Rosenvingea sp. or on the other hand dangerous pad, Turbinaria or spiked leaf are eaten as soup or omelet Colpomenia sp. or on the other hand papery ocean bubble as hack soup, stew, or plate of mixed greens.

Hydroclatharus sp. or on the other hand ocean colander, Dictyota sp. or on the other hand brown, Padina or ocean fan strip weeds as a food dressing, soup, or stew (Novaczek, 2001). Another brown seaweed species, Hydroclathrus spp., can likewise be reaped from the wild, yet being occasional.

In mild territories like Japan, the species just happens during the warm water of spring (Kogame, 1993), while it happens throughout the late spring in the tropics like that of the Philippines (Trono, 1998).

In the last mentioned, the species regularly happens in sprout, shaping thick tangles on reef pads. Current data on the mass culture of the species isn't yet accessible in light of the fact that business utilization of Hydroclathrus mixes is yet to be created (Trono, 2001). All things considered, Marquez et al. (2014) proposed the capability of Hydroclathrus to create biomass.

2.2.3. Green Seaweeds

Green seaweeds are found on both sandy and rough seashores. Many can endure low saltiness and will colonize zones where streams meet the ocean. The green shade of the seaweed is because of the green color chlorophyll required for the photosynthesis of light (Figure 2.3).

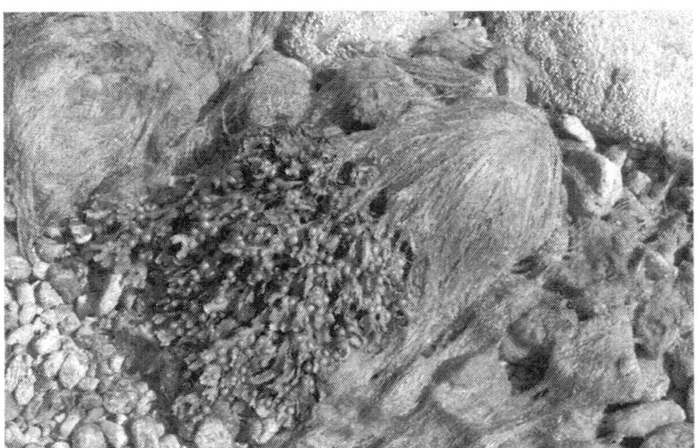

Figure 2.3. Green seaweed.

Source: Image by Wikimedia.

Utilizing just chlorophyll implies that green seaweeds require great degrees of light and thusly won't flourish in shadowed territories or to any profundity. It gives them a preferred position, the capacity to live higher up shore without rivalry from the red or brown seaweeds.

The green seaweeds Ulva sp., Enteromorpha sp., Monostroma sp., Caulerpa sp., Codium sp., are generally known as a wellspring of food. In Asia, nations particularly Japan, dried fronds of consumable Monostroma sp. what's more, Enteromorpha sp are being known as 'aonori-green laver ele-lulua-lumi boso'.

These algae are eaten by people as palatable crude, dried, or cooked. They utilized in preparation of 'nori-jam' soup (Lobban and Harrison, 1994; Novaczek, 2001). In Australia, Paul et al. (2014) revealed comparative biomass yield esteems from similar species refined in raceway.

Also, green seaweeds are known to deliver broad blossoms in eutrophic waters. For instance, Ulva spp. had a biomass of 2.60 kg wet wt/m^2 during a sprout in Mactan, Cebu (Largo et al., 2004).

In the notorious sprout in Brittany, France, and Qingdao, China, the complete biomass of Ulva that was gathered reached as much as 100,000 tons (Smetacek and Zingone, 2013) and more than 1,000,000 tons (Liu et al., 2013), separately. A few types of green algae are additionally viewed as intrusive; among the most infamous is Caulerpa taxifolia.

In the Mediterranean Sea, the species was accounted for to have up to 518 g dry wt/m^2. Codium delicate, additionally considered as an infamous intruder, was discovered developing on eelgrass rhizomes in Canada. It was accounted for by Drouin et al. (2012) to have a biomass of ~10 kg/m^2, while those that attacked northern Chile had a biomass of around 22.9 kg/m^2.

2.3. FARMING OF SEAWEEDS

Cultivation of seaweeds at sea is a very new and hearty type of aquaculture. World production has been portrayed by exponential development during the most recent 50 years (Loureiro et al., 2015) and it significantly increased somewhere in the range of 1997 and 2012, from 7 million tons to 24 million tons (Mt) (FAO, 2014).

In that capacity, seaweeds may one day assume a lot greater job in food gracefully. While seaweed cultivating is extending quickly in a couple of nations and gradually picking up acknowledgment in others, it is still a long way from turning into a typical practice in a large portion of the ~150 nations and domains of the world supplied with coasts (Figure 2.4).

Figure 2.4. Seaweed farming.

Source: Image by Wikipedia.

However, the effectively critical increment in the production of and interest for cultivated seaweeds is demonstrative of their potential as a huge supplement to agrarian yields. Other than broad coastal zones, tremendous breadths of ocean may be transformed into ecofriendly developed fields.

Utilizing momentum advancements and culture ideas, such enormous scope production would yield monstrous measures of seaweed biomass in a controlled way for an assortment of employments.

Seaweeds for food, feed, and, potentially, for biofuels just as for an assortment of forte biochemicals can be created without arable land, freshwater, or manure. Besides, seaweed development gives an assortment of biological system administrations, which are significant naturally and monetarily (Neori et al., 2007; Chopin, 2014; Radulovich et al., 2015).

Extension of seaweed cultivating to scales that approach traditional horticulture relies upon the interest for seaweeds and their items, advancement of cultivating techniques that contend in cost and scale, and acknowledgment of the biological administrations' seaweed ranches can give.

But claim to fame and gourmet items, worldwide interest for seaweeds is intently reliant on whether horticulture will fulfill developing food, feed, and biofuel needs, a job that is flawed at present and maybe far-fetched for the future (OECD-FAO, 2012; FAO, 2013b; Garnett et al., 2013).

This codependency, in any case, may move as cultivated seaweed items gain status as food and for different uses on their own benefits. Generally speaking, as is frequently the situation with new exercises, an endless loop restrains the headway of seaweed cultivating and use in many nations, where "there is no flexibly in light of the fact that there is no interest, and there is no interest on the grounds that there is no gracefully."

This inactivity is in effect effectively tended to in numerous pieces of the world as prove by the developing group of writing advancing or portraying seaweed cultivating and use.

However, the need "for guidelines and the foundation of 'best practices' for seaweed reaping, the board and development" has been noted (Rebours et al., 2014), and these must be created and applied.

As the general standards and purposes that decide seaweed cultivating are solidified, a deductive or principle-based approach will reinforce the predominant inductive or case-based methodology, adding to the following period of the business' development.

In this unique circumstance, this part portrays the basics of seaweed cultivating with the target of being both a record of the best in class just as a commitment to the business' economical extension.

2.4. SEAWEED PRODUCTION AND USE

Seaweed use by gathering or gathering from the indigenous habitat, wild reap, including seashore cast assortment, is an antiquated human practice in numerous areas of the world, similar to China (Tseng, 1981; Tseng and Chang, 1984; Xia and Abbott, 1987) and Europe (Kain and Dawes, 1987; Critchley and Ohno, 1998; Critchley et al., 2006).

There is proof of seaweed used for food and medication from more than 14,000 years prior to southern Chile (Dillehay et al., 2008). Reed (1907) noticed that "numerous huge amounts of these seaweeds are assembled and eaten" in Hawaii, and depicted manners by which they were transplanted starting with one spot then onto the next.

Cultivating of seaweeds started to develop in the mid-twentieth century, especially in Japan and China, outperforming wild collect in tonnage around 40 years back. Refined seaweeds are said to have lightened hunger during China's 3 years (1959–1961) of starvation (Neushul and Wang, 2000).

Seaweeds, or marine macroalgae, are plant-like living beings grouped into three phyla of red, brown, and green, with an all-out number of species evaluated at somewhere in the range of 8,000 and 10,500 (Lüning, 1990;

Thomas, 2002; Hurd et al., 2014; Guiry, 2015), despite the fact that these figures are referred to with no supporting proof.

Seaweeds are universal in coastal zones, showing broad provincial species extravagance and worldwide decent variety designs (Abbott and Norris, 1985; Kerswell, 2006). Zemke-White and Ohno (1999) archived 145 species utilized for human utilization and 101 for hydrocolloid production. FAO (2013) has detailed the huge development of 34 species around the globe.

Cultivating of extra species is usually detailed in developing numbers however frequently in decreased scale or starter structure (e.g., Critchley and Ohno, 1998; Critchley et al., 2006; Hayashi et al., 2014; Radulovich et al., 2015), yet growing the potential.

For instance, an assortment of Porphyra vietnamensis, from the shorelines of India, which is a tropical type of a variety regularly considered of cold waters, has been proposed for tropical development as it develops well at 32°C (Sahoo et al., 2006).

The bounty of local seaweed species in most coastal areas and preparatory standards weigh against specific focal points of advancing the utilization of demonstrated nonnative species.

The presentation of Kappaphycus and Eucheuma cultivating for carrageenan all through the tropics has represented a significant part of the development of the seaweed area in late decades (Bindu and Levine, 2011; Valderrama et al., 2013; Areces et al., 2014).

Scattering these high-yielding species, whose items have a setup showcase, has served to advance seaweed cultivating in any event 30 diverse tropical conditions from Asia to Africa, Latin America, and the Pacific Islands (Valderrama et al., 2013; Areces et al., 2014).

In spite of the fact that the obtrusiveness of these species has been demonstrated to be restricted (Bindu and Levine, 2011; Areces et al., 2014; Sellers et al., 2015), its spread can happen on delicate corals (Bagla, 2008; Bindu and Levine, 2011; Sellers et al., 2015).

In a few cases presentations were led without endorsement from neighborhood administrative bodies (Pickering et al., 2007; Areces et al., 2014; Sellers et al., 2015) or happened inadvertently by transportation (Pickering et al., 2007).

Issues identified with the presentation of nonnative species compound with a summed-up absence of sufficient strategies and guidelines for seaweed cultivating in many nations and require brief consideration.

Disregarding copious seaweed biodiversity, just six genera and a couple of more animal types gave 83.9% of world production in 2012. Eucheuma spp., Kappaphycus alvarezii are fundamentally developed for phycocolloids in tropical waters, while Japanese kelp, Saccharina japonica, once in the past Laminaria japonica, when the main single species by weight in world aquaculture (Neori and Nobre, 2012), presently outperformed by Eucheuma spp.-is developed for the most part for food in mild waters.

While similar genera have stayed prevailing, significant changes since 2000, from information introduced by Sahoo and Yarish (2005), are that Eucheuma and Kappaphycus moved to initially put from fourth and Gracilaria from fifth to third.

Additionally, yields were expanded 3.5 occasions in that period. The predominance of just a couple of seaweed genera and species in culture is comparative additionally in traditional horticulture. Under 20 types of grains, beats, sugar yields, and roots and tubers speak to over 70% of world horticultural yield, also the modest number of cultivated creature species (FAOSTAT, 2015).

Generally speaking, red seaweeds speak to 54.4% of revealed cultivated tonnage around the world, yet include nearly the whole yield of tropical cultivating. Brown colored seaweeds speak to 33.6%, green just 0.08%, and random or unclassified, as revealed especially for China, 11.9% (FAOSTAT, 2015).

While 33 nations detailed seaweed cultivating in 2012 for a sum of 23.8 Mt, 98.7% of production originated from these eight Asian nations, arranged by amount: China (53.6%), Indonesia (27.4%), the Philippines, Republic of Korea, Japan, Korea DPR, Malaysia, and Vietnam.

Zanzibar in Tanzania involves the ninth spot with 0.63% of world production. Along these lines, nine nations (four calm and five tropical) delivered 99.3% of cultivated seaweed yield. Collecting from the wild has stayed stale at ~1 Mt/year for the most recent decade, with 1.1 Mt in 2012 or 4% of all out worldwide seaweed production (FAO, 2014; FAOSTAT, 2015).

Probably, there is more seaweed collecting from the wild with nearby, unreported use. Subsequently, all out cultivated and wild-gathered yield of

~25 Mt in 2012, however sizable, is still just ~0.3% of the all-out plant food created in 2012 through agribusiness.

Notwithstanding, this development is now proof that an enormous extension in seaweed cultivating is practical both in fact and financially in different conditions.

A financial investigation of seaweed development around the tropical world has demonstrated that in certain nations, fundamentally Indonesia and the Philippines, Kappaphycus cultivating is as of now an entrenched industry, giving work chances to a huge number of coastal occupants who have decreased access to elective business sources (Valderrama et al., 2013, 2015).

While in mild nation most production has generally been for direct use as food, in tropical countries the greater part of seaweed cultivating is later and for phycocolloid production, fundamentally Eucheuma spp. also, K. alvarezii, with 33.3% of world production.

In Indonesia seaweed cultivating developed from 0.2 Mt in 2003 to 6.5 Mt in 2012, an extension of multiple times in only 10 years (FAOSTAT, 2015), showing the possibility to extend development in a brief timeframe where conditions are reasonable.

As indicated by FAO (2014), ~9 Mt, 36% of world production, were utilized legitimately for food in 2012. Chopin (2012), in any case, thinks about that 19 Mt, ~75% of world production, was utilized as food while just 11% of the tonnage served for phycocolloid production.

However, in a few tropical Asian nations seaweed utilization as food is omnipresent, counting, e.g., a well-known brand of potato chips prepared with seaweeds. Despite the fact that food production isn't as of now the fundamental motivation behind seaweed cultivating in these locales, a move may have happened as food needs develop and downstream preparing progresses.

2.5. SEAWEED CULTIVATION TECHNIQUES

Control is basic in both agribusiness and aquaculture, including seaweed cultivating. Albeit all out control isn't feasible even in the most refined development chambers and aquaria, there are critical control-based contrasts among cultivating and regular development.

As previously demonstrated, as a result of the overwhelming idea of a few of the factors included including biotic stressors and especially factors identified with water and atmosphere, a significant number of the control methodologies must be imbedded in the seaweed cultivating framework before it is set up.

In that sense, similarly as any effective horticulture may be "atmosphere and soil-keen," fruitful seaweed cultivating must be "atmosphere and water-shrewd." For this, legitimate choice of site, seaweed species and spread, and development strategies are fundamental explicit components that decide control for every circumstance.

2.6. TRENDS IN CONSUMPTION

By far most of the seaweeds gathered and refined are devoured by people (Loureiro et al., 2015), either as an immediate food source, for example, "wakame" from Undaria pinnatifida and "kombu" from S. japonica (once in the past Laminaria japonica), or as a food added substance, for example, carrageenan from Eucheuma spp. what's more, mannitol, alginate, and iodine from S. japonica.

There has been a consistent increment in the utilization of S. japonica since revealing began. Since the cultivating of Eucheuma began decisively in the late 1990s, there has been an exponential development in utilization of this seaweed, while the other principle genera have encountered littler yet noteworthy increments in utilization.

Lately, look into on the medical advantages of seaweeds has indicated that they contain a scope of bioactive aggravates that play out a significant job in tweaking incessant ailment.

The advantages to human welfare from seaweeds incorporate cell reinforcement, anticoagulant, antithrombotic, calming, antitumor/antiproliferation/anticancer/antimetastatic, antivirus, inhibitory impact on parasites, antidepression impacts, immunostimulatory impacts, defensive consequences for the sensory system, and helpful impacts in medical procedure and cerebrum injury (White et al., 2014).

2.7. ECONOMIC RELEVANCE OF SEAWEEDS

Precise information on the monetary significance of seaweed across divisions is hard to track down. Shockingly, the FAO database doesn't contain the dollar estimation of seaweed production over the different kinds of seaweed and seaweed item.

They just record send out dollars, and since a significant part of the seaweed delivered will be devoured locally, this gives just a brief look at the worth. To decide an inexact monetary incentive for the primary five genera delivered, we decided the scope of discount costs for dried seaweed and changed over the wet load to DW, utilizing values from the writing.

This gives a gauge of the discount estimation of the worldwide yearly seaweed production of somewhere in the range of US$10.1 and US$16.1 billion out of 2012.

2.8. PROPAGULE (SEEDLING) PRODUCTION

The ability to create propagules of the ideal species or cultivar in adequate number and quality at whatever point required are key angles for any cultivating activity. This is especially significant with respect to some seaweed species or areas on the grounds that copious material for vegetative engendering isn't generally accessible and sexual or a biogenetic reproduction is regularly mind boggling and expects ability to be actualized.

For instance, Lin et al. (2008) indicated that the green seaweed Enteromorpha proliferate (presently Ulva proliferate) has seven pathways of reproduction including sexual, a biogenetic, and vegetative spread.

Confounding the issue, morphological viewpoints inside every specie, including comparative or unique gametophyte and sporophyte stages, make it important to decide or choose the best possible morphology to cultivate (e.g., Druehl, 2001; Hurd et al., 2014).

Sahoo et al. (2002) gave an intriguing case of this for Porphyra, showing that in 1949 its sporophyte (conchocelis stage), previously viewed as an autonomous living being, was coordinated with its gametophyte and that this finding totally reformed the business (Figure 2.5).

Figure 2.5. Propagule production.

Source: Image by PxHere.

Because of adaptational conduct, numerous seaweed species normally will in general break down absolutely or for the most part after reproduction or in light of regular or other exceptional or fast changes.

As models, going up against the cold of winter may harm the harvest and the shortening of day length may trigger unfortunate regenerative changes (Vásquez, 1995), and Russell (1986) depicted the way toward choosing barren plants of the red seaweed Chondrus crispus, which "will in general break down when they arrive at conceptive development," as the reason for creating cultivars for cultivating.

In the event that unaccounted for, this sort of conduct may speak to an obstruction to cultivating, for spread as well as far as getting yields. In view of this, for the fundamental mild zone seaweeds, Saccharina, Undaria, and Pyropia, proliferation is through spore development (Hurd et al., 2014).

For subtropical and tropical cultivated species, pieces of thalli are regularly adequate as vegetative propagules.

While all of Eucheuma and Kappaphycus production depends on straightforward vegetative spread acquired from the past gather (Teitelbaum, 2003; Valderrama et al., 2013), for Gracilaria and Sargassum vegetative proliferation is the most widely recognized technique, despite the fact that they are likewise imitated by spore development (Redmond et al., 2014a, b).

2.9. POTENTIAL SEAWEEDS FOR BIOFUEL PRODUCTION

The different employments of seaweeds and their side-effects have become indispensable pieces of human everyday life. Seaweeds are most broadly used as food, establishing a significant piece of the human eating regimen, especially among coastal populaces and in numerous Asian nations, for example, China, Japan, and Korea.

The phycocolloids created via seaweeds are additionally generally utilized as settling, thickening, suspending, and gelling specialists in nourishments, organic product drinks, syrups, milk-based items, toothpaste, beautifying agents, and even in material paints (Prudhomme van Reine, 2001; Sahoo and Yarish, 2005).

Seaweeds are likewise utilized in farming as compost and animal feed just as in the clinical and pharmaceutical businesses. Seaweed biomasses have been widely advanced for contamination bioremediation through mono-or coordinated culture (FAO, 2009; Fei, 2004; Neori et al., 2004) and for supportable feedstock for biofuel production (Gao and McKinley, 1994; Prudhomme van Reine, 2001).

Seaweeds are customarily gathered from nature; however, the expanding interest for their biomass frequently brings about overharvesting and exhaustion of normal stocks.

As a reaction, normalized, schedule, and prudent culture strategies were created for different economically significant seaweeds, for example, Pyropia/Porphyra, Saccharina (Laminaria), Undaria, Gracilaria, Kappaphycus, and Eucheuma (Sahoo and Yarish, 2005).

The FAO (2014a, b) announced in 2012 that roughly 20.9 million tons (wet weight) of seaweeds were all around delivered through aquaculture. Of these, the red seaweeds commanded the biomass gather with about 12.91 million tons, trailed by the brown seaweeds with 7.96 million tons and green seaweeds with roughly 0.02 million tons.

Besides, Marquez et al. (2014) featured a couple of standards to distinguish seaweed species that can be utilized to deliver biofuel. In that, the creators demonstrated that predictable and simple access just as a high and consistent gracefully of seaweed biomass that will be utilized as feedstock for biofuel are essential to economically reduce, if not totally supplant, the reliance on petroleum derivative determined vitality.

Manageable editing and accessibility of the strategy for mass culture of the seaweed species are among the essential concerns. The way of life strategies for the previously mentioned seaweed species can be tapped to build the biomass production for biofuel.

In this, we feature some potential seaweed species from various areas that can be utilized as feedstock for biofuel. Our rundown is especially centered around species that can deliver high biomasses as well as those that can be or are being exposed to mass development.

2.10. BASIC POSTHARVEST HANDLING

Postharvest taking care of has the motivation behind rendering the collected seaweed item perfect and prepared for the expected use or for capacity. Contingent upon a few elements, dealing with can change generously and ought to be done simply in the wake of confirming that the reaped item is liberated from contaminants, including fecal issue, oil, red tides, and different contaminations that might be or have been available in seawater.

Tidiness of developed seaweeds is something that ought to likewise be evaluated previously or during reaping. With a recurrence and expansiveness reliant on proposed use and site attributes, eventually during postharvest taking care of, tests of the gathered material ought to be exposed to microbiological and chemical investigation to recognize potential contaminants just as wanted chemical and dietary qualities.

In the wake of cleaning, drying is the most well-known postharvest treatment, in spite of the fact that seaweeds may likewise be devoured or handled new.

A significant number of the components that decide essential postharvest taking care of strategies identify with the species being developed, the utilization of the item and downstream procedures, and the scale and specialized degree of the activity, which additionally to a great extent relates with accessible offices ashore.

A significant thought is whether the item will be utilized for crude mechanical handling (e.g., phycocolloid extraction), which is less requesting of value and consistency, or for food or the extraction of refined biochemicals, which require the most extreme consideration.

The size of postharvest tasks changes from modest quantities of seaweed newly reaped from waterfront plantings for direct home or neighborhood utilization to activities that handle numerous tons every day and accordingly require tough procedures to ensure quality.

Despite the fact that preparing adrift has been proposed (e.g., Notoya, 2010), practically all postharvest dealing within current frameworks are done ashore. After gathered material is brought to shore, pollutions are evacuated, frequently as a major aspect of a primer washing.

Fouling, fauna, tying strings, and an assortment of different polluting influences are expelled by hand, as are undesirable and harmed parts. Seaweeds are then altogether washed, which should be possible incompletely or totally, with seawater, frequently within any event a last freshwater flushing to evacuate salt if so wanted.

During or in the wake of washing the material can be cut or minced relying upon the necessities (e.g., mincing is standard for "nori" sheet planning). The item is then prepared for conservation or for direct new use, which can be utilization or handling before long, or bundling followed by cool stockpiling to safeguard newness.

Working with a few tropical animal categories, Radulovich et al. (2015) found that cleaned and plastic-stowed material from most species held newness through refrigeration for at any rate fourteen days, except for Caulerpa racemose (an assortment of ocean grapes), which kept going just 5 days.

For some uses, uninvolved sun oriented drying by putting seaweeds on surfaces or balancing them from racks is adequate and broadly used to deliver material with ca. 30–35% dampness content. This strategy, although minimal effort requires care of the drying material for a few days, including going it to guarantee that drying is even.

The procedure additionally leaves the item defenseless against tainting or to harm by downpour if not secured and creates variable last dampness levels (e.g., Gideon, 2007; SEAFDEC, 2007; De San, 2012).

Constrained breeze and warmth helped solar driers are being utilized as a stage above latent sun based drying. In bigger or modern tasks, mechanical or dynamic drying is done in passages or stoves with the contribution of warmth from outside sources and constrained air might be utilized to accelerate the procedure.

This controlled procedure permits the treatment of huge volumes of seaweeds while accomplishing a normalized extreme dampness content, which can be lower than levels realistic through sun powered drying (e.g., 20% or less).

Previously or during drying, the material might be taken a shot at, e.g., Saccharina cutting edges are smoothed and compacted, and edges and the base are sliced to acquire the ideal shape and new Undaria fronds are blended in with cinders to save alluring attributes (Hurd et al., 2014).

A high salt substance might be kept up on the outside of the material while drying, with some seaweed items even intentionally salted, so as to upgrade protection. Subsequent to drying, the material is either prepared or bundled/baled and put away in obscurity at room temperature, with no refrigeration required.

Considering that, new seaweed material contains ca. 90% water, drying it helps in safeguarding as well as significantly lessens volume and weight for transport. Other than cool stockpiling and drying, a few different procedures can be utilized to ensure both conservation and item quality, while likewise helping explicit procedures.

For instance, for some phycocolloid-yielding seaweed, especially Sargassum for the creation of alginates, certain chemical protection strategies are utilized.

Dark (1955) and Taizo and Hiroshi (1955) did spearheading research on chemical conservation techniques focused at capturing the development of microscopic organisms and different microorganisms, which corrupt the nature of the seaweed and its item during capacity.

Other essential postharvest dealing with forms identify with methods used to improve quality and innocuity of the item, especially for use as food.

For instance, whitening (drenching the perfect and new seaweed material in bubbling water for a brief timeframe) turns brown and red seaweed produce green (since chloroplasts burst) and probably-cleans it.

Subsequent to whitening, the item might be expended, prepared, dried, or even solidified. Sanitization is another significant essential procedure that can be actualized before utilization as food or before drying.

Drying at certain base temperatures (i.e., somewhere in the range of 60°C and 70°C) can by and large be viewed as a sufficient purification process, yet its viability stays to be resolved for seaweed material.

2.11. ECOLOGICAL AND ENVIRONMENTAL IMPACTS OF SEAWEED FARMING

Seaweed cultivating is an extractive aquaculture whose very procedure of creation of significant biomass renders the ocean's different biological system administrations with natural and monetary qualities (Chopin et al., 2008, 2010; Neori et al., 2007; Radulovich et al., 2015).

When contrasted with animal aquaculture, seaweed cultivating includes oxygen during photosynthesis and cleans seawater from abundance supplements (N, P, and others). Supplement extraction, or take-up, cleans water viably and altogether through a procedure known as bioremediation, with a few reasonable applications (e.g., Forster, 2008; Neori, 2008).

It has additionally been demonstrated that cultivating seaweeds in seaside waters upgrades biodiversity and fisheries (Radulovich et al., 2015). Seaweeds are carbon sinks that can decrease sea fermentation through the take-up of CO_2 from water at a scale that relies upon the size of cultivating tasks.

It has been assessed that to settle environmental CO_2 at 400 ppm, worldwide carbon sequestration should be a few hundred gigatons of carbon, and Hughes et al. (2012) broke down the utilization of seaweeds for carbon-negative fuel creation.

As an idea work out, and with an expected gross seaweed catch pace of 2000 t C/km^2/year, it would take 5 million km^2 (1.4% of the world's sea zone) to take up 10 Gt C/year by sea developed seaweed, generally the absolute yearly worldwide carbon emanation from utilization of petroleum derivatives (Landschützer et al., 2014).

This kind of monstrous biomass creation plan could reasonably reduce the world's food hole and give a wellspring of bioenergy. Large-scale seaweed cultivating can have both positive and negative effects on biodiversity.

Concealing can and alkalization may damage and advantage distinctive nearby organic exercises, going after occurrence with phytoplankton and accordingly channel feeders, and yet supporting calcification of shellfish and corals, which experience the ill effects of sea fermentation (Branch et al., 2013).

The mix of huge scope extractive seaweed cultivating with the extractive cultivating of bivalves and the fed cultivating of fish in confines may permit an environmentally and practically adjusted turn of events (Chopin et al.,

2008, 2010; Neori et al., 2007), however restricted in scale to regions encompassing fish cultivating tasks.

Nonetheless, any movement that occupies room adrift in a lasting way additionally raises concerns. While the cultivating of nonnative species was talked about before, seaweed cultivating can infringe navigational and recreational territories, make stylish visual irritation, block fisheries, and strife with different employments of ocean zones.

Actually, since seaweed cultivating is for the most part a territory-based undertaking so as to catch insolation, its belongings might be more articulated than, e.g., confined cultivating or suspended bivalve cultivating, which are from multiple points of view three-dimensional undertakings.

In certain circumstances, especially close to shore with exceptionally shallow waters, lowered cultivating is conceivable, yet in more profound waters, surface cultivating is the standard-with the preferred position this typically is past quick waterfront waters and in this way struggle with other water utilizes is insignificant.

Seaweed cultivating in both national and universal waters requires another lawful structure that secures the interests all things considered, including cultivating business people (Rosenthal et al., 2012; Wever et al., 2015).

2.12. ECONOMIC AND SOCIAL CONSIDERATIONS OF SEAWEED FARMING

Financial contemplations change, relying upon elements, for example, the species cultivated, the motivation behind cultivating, showcase costs, and the size of activities. The way that seaweed cultivating proceeds and extends, in most, if not all, cases without contortions, for example, sponsorships, verifies its gainfulness.

A few examinations give important proof on the monetary advantages of seaweed aquaculture improvement in both calm and tropical districts.

For instance, Edwards, and Watson (2011) led a monetary examination of Laminaria digitata cultivating in Ireland; Philippsen et al. (2014) introduced a cost examination for ethanol delivered from cultivated seaweeds.

SINTEF (2014) portrayed another bioeconomy for Norway dependent on the development and handling of seaweeds; Lenstra et al. (2011) and

Van lair Burg et al. (2013) analyzed the monetary achievability of seaward seaweed creation in the North Sea.

Msuya (2011) considered the effect of seaweed cultivating on the financial status of beachfront networks in Zanzibar; and Valderrama et al. (2013, 2015) broke down the financial matters of cultivating for carrageenan in tropical conditions, as talked about in more noteworthy detail for the rest of this area.

2.13. POSSIBLE INDUSTRIAL APPLICATIONS OF SEAWEEDS

2.13.1. Pharmaceuticals

Throughout the years, there are noteworthy advancements in medicate plan and profoundly dynamic chemical mixes can be orchestrated. However, regular items are considered as incredible hotspots for the advancement of new pharmaceuticals which are increasingly powerful with less unfriendly impacts.

Subsequently, utilization of characteristic items as ailment cures has a long history, and common items examine keeps on giving a gigantic assortment of lead structures which are utilized as formats for the advancement of new medications by the pharmaceutical business (Borris, 1996).

As of late, marine regular items have been perceived as viable medication applicants. In such manner, Seaweeds, and its subsidiaries may be utilized as helpful normal items for the improvement of compelling pharmaceuticals. In any case, further examinations are required before clinical application and commercialization of seaweeds and its subsidiaries as medications.

2.13.2. Functional Foods

Practical nourishments are one of the top patterns in the food business. They give more than essential sustenance. They have bioactive food segments that can conceivably improve wellbeing when eaten all the time as a component of a shifted diet.

Utilitarian nourishments are a developing field of food science because of their expanding fame with wellbeing conscious shoppers, and subsequently, there is a developing enthusiasm among makers just as people in general in practical nourishments or food sources that may give medical advantages past essential sustenance (Spence, 2006; Marinangeli and Jones, 2010).

Epidemiological examinations and clinical preliminaries completed in various nations have shown and proposed that practical nourishments ordinarily incorporate welfare cases.

As an example of that, decrease of malignant growth chance, improvement in heart wellbeing, incitement of safe framework, improvement in gastrointestinal wellbeing, upkeep of urinary tract wellbeing, mitigating impacts, decrease of hypertension, support of vision, antibacterial, and antiviral exercises, and anti-obese impacts identified with the specific useful fixings.

The flow correspondence explains the key achievement of brown seaweeds affecting the improvement of practical nourishments or dietary enhancements regarding the organic exercises of optional metabolites.

2.13.3. Cosmeceuticals

Cosmeceuticals are topically applied class of corrective items professed to have organically dynamic fixings with therapeutic or medication like advantages (Gao et al., 2008). By and large, they fulfill the requirements of magnificence and soundness of skin, hair, or nails.

At present, quickly developing number of cosmeceutical items asserting to assurance, brightening, or depigmenting, hostile to wrinkling and against maturing decline redness and saturating. Throughout the years, apparently enthusiasm for normal items has been expanded.

Conventional cosmeceuticals include the topical use of naturally dynamic fixings that influence the skin wellbeing. In a perfect world, improvement in practical properties of cosmeceuticals utilizing characteristic fixings would upgrade the key advancement patterns and new item improvement open doors too.

The practical fixings in topically applied cosmeceuticals assume a crucial job in healthy skin. Along these lines, the push towards regular fixings ought to have progressively logical based course (Nohynek et al., 2010).

Now, the notable practical properties of Seaweeds, for example, cancer prevention agent, calming, antiallergy, defensive impact against UVB-incited oxidative pressure and MMP inhibitory movement have given promising logical base to create healthy skin cosmeceuticals.

2.14. SEAWEED EXTRACTS IN AGRICULTURE

In natural farming and agriculture, weakened concentrates of seaweed are applied to advance development, forestall bugs and ailments, and improve the nature of the items. The viability of the concentrates is likely founded on plant hormones (principally cytokinins (CKs)) and follows supplements present in the concentrates.

Ocean and land have their own characteristics, unmistakable from one another. The ocean is in numerous regards unfriendly to the land, because of its high substance of salt (NaCl), which is harmful to vegetation on the land.

Anyway, the ocean is wealthy in supplements, significant supplements and follows supplements. These supplements, in blend with the vitality of the sun, structure the reason for plant and creature life in the ocean.

Correspondingly, plants can't develop on the land without being supported by these supplements. In soils kept by the ocean (marine sands and marine dirt) the major and follow, supplements have their birthplace in the ocean; one may state that the ocean quality has been safeguarded in this dirt. Numerous dirt is poor in large scale and micronutrients. For example, sandy soils which were framed by spread sands stored during the Pleistocene and which involve broad regions in the Netherlands, the bordering Rhine zone, Belgium, and Northern France are especially poor in supplements.

Supplements beginning from the ocean, packed in seaweeds up to multiple times, may be appropriate to enhance these poor sandy soils. In agribusiness and cultivation, seaweed has discovered application as creature food, as soil conditioner and fertilizer, and, as fluid concentrates, as development advertiser and yield protectant against bugs and maladies.

Verifiable records show that the utilization of seaweeds in farming is old and broad any place there are plentiful supplies. In the beachfront areas of Iceland, Norway, Great Britain, Ireland, and France creatures still normally feed upon new seaweed or are given arranged seaweed food.

As a dirt conditioner, certain red seaweeds, referred to at the French coast as 'maerl,' are utilized in corrosive humus-rich soils or peat soils as a choice to lime, as a result of their very high (up to 80%) substance of calcium carbonate (Blunden, 1991; Chapman and Chapman, 1980).

2.15. PROPERTIES OF SEAWEED EXTRACTS

2.15.1. Species of Seaweed

Seaweed removes are produced using storm-cast or newly cut seaweeds. The overwhelming species in the North Atlantic Ocean utilized for making removes is Ascophyllum nodosum, having a place with the brown-green growth.

Different species utilized, all having a place with the Phaeophyceae are Laminaria hyperborea and L. digitata, and Fucus vesiculosus and F. serratus. In the Southern half of the globe, the brown-green growth Ecklonia maxima and Durvillea potatorum are utilized for making removes.

2.15.2. Quality of Seaweed

Shockingly little consideration has been given to the nature of the seaweed shaping the premise of the concentrates. It glances clear in the arrangement of seaweed concentrates to begin from newly cut rather than storm-cast seaweeds, however, it is indistinct how far makers of seaweed extricate recognize the centrality of this angle.

In like manner, data is rare about which seaweed species are the most appropriate for making extricates, at which development stage the seaweed ought to be reaped, which parts of the seaweed ought to be utilized, and when the seaweed ought to be collected in the developing season.

As will be talked about beneath, follow supplements and plant hormones, specifically CKs, are the dynamic constituents in seaweed separates. Since youthful developing fronds ('leaves') of seaweeds likely contain the most noteworthy measures of CKs and follow supplements, they can best be gathered in the spring.

Notwithstanding, in pre-winter, the fronds are rich in polyphenols with hostile to parasitic exercises (Kirkham and Flood, 1956), so the estimation of concentrates from seaweed collected in the pre-winter should likewise be thought of.

2.16. CONCLUSION

Seaweeds have low substance of lipids but they are of high caliber as far as dietary benefit. Nature of protein and lipids are as adequate as those of some other dietary vegetables because of high substance of fundamental amino acids and moderately more elevated levels of unsaturated fats present in ocean growth.

Seaweed is brought directly into the field in beachfront locales and it is applied as dried and ground seaweed supper.

When expended, they additionally have a significant job in the change of lipid digestion in human body. In other words, seaweeds can be considered as modernly encouraging plants shaping one of the significant marine living assets of high healthy benefit.

This chapter effectively describes the entire essential; factors related to the seaweeds, their sources, and their uses in the industry as well. It evaluates the agricultural aspect of seaweeds and their properties as well.

REFERENCES

1. Kennedy, J., (2019). What are the 3 Types of Sea Weed (Marine Algae)? [online] ThoughtCo. Available at: https://www.thoughtco.com/types-of-marine-algae-2291975 (accessed on 10 July 2020).
2. Kılınç, B., Cirik, S., Turan, G., Tekogul, H., & Koru, E., (2013). Seaweeds for food and industrial applications. In: Food industry. IntechOpen.
3. Kumar, C. S., Ganesan, P., Suresh, P. V., & Bhaskar, N., (2008). Seaweeds as a source of nutritionally beneficial compounds-a review. Journal of Food Science and Technology, 45(1), 1.
4. Renn, D., (1997). Biotechnology and the red seaweed polysaccharide industry: Status, needs, and prospects. Trends in Biotechnology, 15(1), 9–14.
5. Tiwari, B. K., & Troy, D. J., (2015). Seaweed sustainability-food and nonfood applications. In: Seaweed Sustainability (pp. 1–6). Academic Press.

CHAPTER 3

Nutritional Composition of Seaweeds

CONTENTS

Seaweeds have been part of consumption in the Asian continent for a long time. It has been part of their dietary food. In some events food prepared using seaweeds were provided to the special people and kings of ancient times. This chapter describes the nutritional composition of seaweeds. The characteristics such as proximate composition and vitamins present in them have been evaluated at first. Their use in the dietary supplement has also been defined along with the nutritional values of red, green, and brown seaweeds. The importance of seaweeds is also found in the medicinal purposes and hence it has been concluded that in order to find their true value in the future medicines a wide range of research is required to be performed on them.

3.1. INTRODUCTION

In Asian Pacific, people have a convention of consuming seaweeds as a component of their eating routine while, in the Western nations, the chief employments of seaweeds are as sources of phycocolloids, thickening, and gelling specialists for different mechanical applications utilizes for nourishments (Abbott, 1996; Darcy-Vrillon, 1993; Mabeau and Fleurence, 1993).

As of late, in France, seaweeds have been approved as vegetables and sauces (Mabeau, 1989). In this way, seaweeds have gotten a significant vegetable and a significant food preservative in the human eating routine.

The dietary properties of seaweeds are not totally known at this point, and they are typically evaluated from their synthetic organization alone (Darcy-Vrillon, 1993; Mabeau and Fleurence, 1993).

Contrasted with land plants, the compound organization of seaweeds has been ineffectively researched and the greater part of the accessible data just arrangements with conventional Japanese seaweeds (Fujiwara Arasaki, Mino, and Kuroda, 1984; Nisizawa, Noda, Kikuchi, and Watanabe, 1987; Watanabe and Nisizawa, 1984).

The substance synthesis of seaweeds shifts with species, territories, development, and ecological conditions (Ito and Hori, 1989). By and by, as a rule, seaweeds are rich in non-starch polysaccharides, minerals, and nutrients (Darcy-Vrillon, 1993; Mabeau and Fleurence, 1993).

As seaweed polysaccharides can't be totally processed by human intestinal proteins, they are viewed as another source of dietary fiber and food ingredients (Lahaye, 1991; Mabeau, Cavaloc, Fleurence, and Lahaye, 1992; Mabeau and Fleurence, 1993).

Along with their low lipid content, seaweeds just give an exceptionally low measure of vitality (Jurkovic, Kolb, and Colic, 1995). Utilization of seaweeds can expand the admission of dietary fiber and lower the event of some incessant maladies (diabetes, stoutness, heart illnesses, malignancies, and so forth), which are related with low fiber diets of the Western nations (Southgate, 1990).

Dietary fiber can be isolated into solvent and insoluble parts. The thickness of dissolvable dietary fiber is liable for more slow assimilation and retention of supplements, and lower levels of blood cholesterol and glucose.

Interestingly, insoluble dietary fiber is described by its capacity to increment fecal mass and abatement intestinal travel time (Baghurst and Record, 1996; Potty, 1996). In this way, the physiological impacts of dietary fiber are associated with their specific physico-synthetic properties (Roehring, 1988).

As dietary fiber is the significant segment of the seaweeds, deciding the physico-compound properties of seaweed dietary fiber can give us some comprehension of the physiological impacts of expending seaweeds and their potential applications as texturizing and building specialists in making low-calorie nourishments.

Despite the fact that the seaweed-oras in Hong Kong are genuinely rich, they are generally under-used (Hodgkiss and Lee, 1983). All in all, most Hong Kong seaweeds are chiefly utilized as creature feeds or manures by the beachfront residents (Hodgkiss and Lee, 1983).

Two red seaweeds (Hypnea charoides and Hypnea japonica) and one green seaweed (Ulva lactuca) are inexhaustible in Hong Kong (Hodgkiss and Lee, 1983). The point of the current investigation in this part was to decide the synthetic creation and amino corrosive profiles of these three subtropical seaweeds so as to give progressively complete supplement data about them.

Moreover, some physico-compound properties of these seaweeds were additionally explored so as to assess their latent capacity use as ingredients of food.

3.2. PROXIMATE COMPOSITION

Proximate composition, complete dietary fiber content and physicochemical properties of three brown and two red eatable Spanish seaweeds, in particular:

Himanthalia elongata (ocean spaghetti), Bifurcaria bifurcata, Laminaria saccharina (sweet kombu), Mastocarpus stellatus and Gigartina pistillata were contemplated. Cinders (24.9–36.4%) were high in all examples.

Protein content went from 10.9 to 25.7%, being a lot higher for Laminaria (25.7%) trailed by the red seaweeds (15.5–21.3%). Minor parts were lipids (0.3–0.9%) in all examples with the exception of Bifurcaria (5.6%).

In end, these seaweeds can be evaluated as a decent wellspring of food fiber, protein, and minerals for human utilization (Gómez-Ordóñez et al., 2010). Mineral substance was resolved in a few brown (Fucus vesiculosus, Laminaria digitata, Undaria pinnatifida) and red (Chondrus crispus, Porphyra tenera) edible marine ocean vegetables.

Seaweeds contained high extents of debris (21.1–39.3%) and sulfate (1.3–5.9%). In brown-green growth, debris content (30.1–39.3%) was higher than in red-green growth (20.6–21.1%). Palatable brown and red seaweeds could be utilized as a food supplement to help meet the suggested every day admission of some fundamental minerals and follow components (Rupérez, 2002).

Ocean spaghetti (Himanthalia elongata), Wakame (Undaria pinnatifida), and Nori (Porphyra umbilicalis), on unsaturated fat organization, amino corrosive profile, protein score, mineral substance, and cancer prevention agent limit in low-salt meat emulsion model frameworks.

The expansion of seaweeds caused an expansion in ω-3 polyunsaturated fats (PUFA) and a lessening in the ω-6/ω-3 PUFA proportion. By and large, expansion of seaweeds to items expanded the convergences of K, Ca, Mg, and Mn.

The nearness of Nori caused an expansion in levels of serine, glycine, alanine, valine, tyrosine, phenylalanine, and arginine, while Wakame and Sea Spaghetti created no huge changes in amino corrosive profiles in the model frameworks (López-López et al., 2009).

The healthful pieces of 34 eatable seaweed results of the Laminaria sp., Undaria pinnatifida, Hizikia fusiforme, and Porphyra sp. assortments were analyzed. The marine macroalgae assortments tried exhibited low lipid substance with 2.3 ± 1.6 g/100 g semi-dry example weight (s.w.) and end up being a rich wellspring of dietary fiber (46.2 ± 8.0 g/100 g s.w).

The unadulterated protein substance of seaweed items changed generally (26.6 ± 6.3 g/100 g s.w. in red-green growth assortments and 12.9 ± 6.2 g/100 g s.w. in brown-green growth varieties).

All basic amino acids were identified in the seaweed species tried and red-green growth species highlighted remarkably high centralizations of taurine when contrasted with brown-green growth assortments (Dawczynski et al., 2007).

The total lipid, protein, debris, and individual unsaturated fat substance of palatable seaweeds that had been canned (Saccorhiza polyschides and Himanthalia elongata) or dried (H. elongata, Laminaria ochroleuca, Undaria pinnatifida, Palmaria sp. also, Porphyra sp.) Total lipid content went from 0.70±0.09 to 1.80±0.14 g/(100 g dry weight (DW)).

The four most copious unsaturated fats were C16:0, C18:1ω9, C20:46 and C20:5ω3. Unsaturated fats prevailed in all the Brown seaweeds considered, and soaked unsaturated fats in the red seaweeds; however, the two gatherings are adjusted wellsprings of ω3 and ω6 acids.

Debris content extended from 19.07±0.61 to 34.00±0.11 g/(100 g DW), and protein content from 5.46±0.16 to 24.11±1.03 g/(100 g DW) (Sanchez-Machado, et al., 2004).

3.2.1. Total Dietary Fiber Analysis

The substance of total dietary fiber in seaweeds was resolved by the AOAC enzymatic gravimetric technique (AOAC, 1995). In a nutshell, aliquots of tests (1 g of dry issue) were first rewarded with two amylases, a warmth stable an amylase for 30 min in a bubbling water shower and a contagious amyloglucosidase for 30 min at C to expel starch and afterward a bacterial protease to solubilize protein.

The amylase proteins utilized had been tried to be liberated from glucanase. The compound rewarded blend containing the support arrangement and non-edible materials was encouraged with four volumes of total ethanol.

At that point, the ethanol-insoluble build-up was sifted with a Fiber-Tec System (Tecator 1023, Hoganas, Sweden). The build-up recuperated was washed, stove dried and weighed to give the gravimetric yield of the seaweed fiber material or TDF. The heaviness of TDF was rectified for debris and remaining protein content and a clear.

3.3. NUTRITIONAL VALUE OF SEAWEED

Bioactive mixes are particles from engineered or normal sources that have been organically tested for exercises in various key zones.

The movement of these bioactive mixes has been connected to acceptable security for a long time, and apparently, bioactive food segments can modify the hereditary articulation of a large group of cell occasions, in this way affecting security results or giving useful cancer prevention agent or enzyme inhibitory exercises. Subsequently, these mixes are getting comfortable segments of the standard jargon. This expanded mindfulness is in part upheld by the business center around wellbeing and health just as general wellbeing drives advanced by legislative associations.

The current article measures the nourishing effect of seaweeds as a wellspring of basic supplements just as their key bioactive atoms and exercises; it additionally depicts the individual advantages attributed to the principle sorts of consumable seaweeds.

Business wellbeing cases can here and there be deluding, yet structures are developing to help with explaining the messages transmitted to buyers.

For instance, in the United Kingdom, there is developing open awareness with respect to the wellbeing effect of expending vegetables and natural products, as advanced in the 'five-a-day' general wellbeing efforts, which support a new organic product or vegetable utilization.

Reference supplement admissions have been built up by the UK government as far as the day by day sums important to gracefully fundamental supplements (Figure 3.1).

Figure 3.1. Black sand seaweed.

Source: Image by Flickr.

These qualities have been developed by non-administrative associations and archived as rule everyday sums, which incorporate suggested fiber admissions. The current audit alludes to the two arrangements of information and contrasts them and the built-up supplement parts of seaweeds.

3.3.1. Polysaccharides and Dietary Fibers

Seaweeds contain a lot of polysaccharides, prominently cell divider auxiliary polysaccharides that are extracted by the hydrocolloid business: alginate from brown seaweeds, carrageenans, and agar from red seaweeds.

Other minor polysaccharides are found in the cell divider: fucoidans (from brown seaweeds), xylans (from certain red and green seaweeds), and ulvans in green seaweeds. Seaweeds additionally contain capacity polysaccharides, quite laminarin (β-1,3-glucan) in brown seaweeds and Floridian starch (amylopectin-like glucan) in red seaweeds.

At the point when confronted with the human intestinal microscopic organisms, the greater part of these polysaccharides (agars, carrageenans, ulvans, and fucoidans), are not processed by people, and in this manner can be viewed as dietary strands (Lahaye et al., 1990, 1991). Water-dissolvable and water-insoluble filaments have been related with various physiological impacts.

Numerous gooey solvent polysaccharides (gelatins, guar gum, and so forth.) have been related with hypocholesterolemic and hypoglycemic impacts, while water-insoluble polysaccharides (cellulose) are fundamentally connected with a reduction in stomach related tract travel time (Southgate, 1990). Among polysaccharides, fucoidans were especially concentrated as they demonstrated fascinating organic exercises (hostile to thrombotic, against coagulant, anticancer, against proliferative, hostile to viral, and hostile to reciprocal operator, mitigating).

These properties open up a wide field of potential remedial applications, some of which are now the subject of licenses concerning prominently the counter coagulant and hostile to thrombotic properties (Charreau et al., 1995; Nasu et al., 1997; Angstwurm et al., 1997).

Concerning xylans and laminarins, they are totally and rapidly corrupted by intestinal microorganisms, alginates are just somewhat debased and lead to a significant creation of short-chain unsaturated fats.

3.3.2. Mineral Content

Seaweeds draw from the ocean an exceptional abundance of mineral components, full-scale components, and follow components. The mineral portion of certain seaweeds represents up to 36% of dry issue.

The brown seaweeds have customarily been utilized for rewarding thyroid goiter (Suzuki et al., 1965). The connection among iodine and the thyroid hormones was built up soon a short time later. In spite of the way that this crude material wellspring of iodine has been deserted (aside from in China), Fucus vesiculosus is as yet enrolled in the European pharmacopeia for its high iodine content. As concerns iodine, Laminaria is the fundamental source as it contains 1500 to 8000 ppm DW (fucals contain 500 to 1000 ppm dry load of iodine). Seaweeds are additionally one of the most significant vegetable wellsprings of calcium.

Calcium substance might be as high as 7% of the evaporate weight in macroalgae and to 25 to 34% in the white seaweed lithotamne. Seaweed utilization may in this manner be useful on account of eager moms, adolescents, and older that all presented to a danger of calcium inadequacy. Regardless of height substance, the linkage of specific minerals with anionic polysaccharides (alginate, agar, or carrageenan) may restrict their retention. For example, the solid partiality of divalent cations (especially calcium) for carboxylic polysaccharides (alginates).

Seaweeds are high in minerals because of their marine living space, and the decent variety of the minerals they assimilate is wide. Significant minerals, for example, calcium, collect in seaweeds at a lot of more elevated levels than in earthbound staples (Figure 3.2).

Figure 3.2. Roasted seaweed in the Korean market.

Source: Image by Flickr.

This is represented in an 8 g segment of Ulva lactuca (ocean lettuce), which gives 260 mg of calcium, approaching approximately 37% of the RNI of calcium for a grown-up male in between 12 and 22. In correlation, a similar bit of cheddar gives only 5% of the RNI.

Sodium and potassium are likewise present at generally significant levels, despite the fact that Na: K proportions are normally beneath 1:5.23 Minerals, for example, iron, and copper are available in seaweeds at more elevated levels than in some notable earthly wellsprings of minerals, for example, meats, and spinach.

For instance, there is increasingly iron in an 8 g serving of dry Palmaria palmata (Dulse/Dillisk) than in 100 g of crude sirloin steak (6.4 mg versus 1.6 mg, respectively).

An examination on the bioavailability of iron from Porphyra spp. discovered that iron assimilation and iron maintenance levels were lower in seaweed-took care of creatures than a control and it was proposed this was expected to the polyphenolic mixes or peptides from incompletely processed protein in the diet.

Copper is available at an elevated level in seaweeds. The reference supplement admission of copper is 1.2 mg per day 25 and an 8 g segment of Undaria pinnatifida (Wakame) contains 14% of the RNI for this mineral.

Brown seaweeds bioaccumulate numerous components and are a decent wellspring of magnesium, copper, iron, and iodine just as other rarer components. An 8 g every day admission of 'Kombu,' as utilized in Asian cooking, contains 65% of the RNI for magnesium. Iodine is a significant supplement in metabolic guideline and development designs and is bounteous in many seaweeds.

Seaweeds have been portrayed as a perfect food-safe normal wellspring of the mineral. Consumption of huge sums could actuate some bothersome impacts at the same time; generally speaking, the impacts of iodine utilization would be helpful.

Follow components, for example, zinc, are available in seaweeds and a portion of these, for example, arsenic, have negative wellbeing impacts.

On account of arsenic, further examination of speciation shows that the sort of arsenic is significant in surveying harmfulness; since numerous kinds of arsenic are not processed, these don't represent a hazard to wellbeing. For most of seaweeds, the degrees of irresistible metals are beneath sanitation restrains normally.

3.3.3. Proteins and Amino Acids

The protein substance of brown seaweeds is commonly little (normal: 5–15% of the DW), though higher protein substance is recorded in green and red seaweeds (on normal 10–30% of the DW).

In some red seaweed, for example, Palmaria palmata (dulse) and Porphyra tenera (nori), proteins can speak to up to 35 and 47% of the dry issue, individually. These levels are equivalent to those found in high-protein vegetables, for example, soybeans (in which proteins speaks to 35% of the dry mass).

The protein levels of Ulva spp. are in the range 15–20% of the DW. Aside from Undaria pinnatifida, which contains 11–24% proteins, other brown seaweeds (Laminaria digitata, Ascophyllum nodosum, Fucus vesiculosus, and Himanthalia elongata) have low protein content. Spirulin, a freshwater microalga, is notable for its high protein content (70% of dry issue).

The edibility in vivo of algal proteins isn't all around archived, and accessible examinations about their digestion by people have not given indisputable outcomes. In any case, a few laborers have portrayed a high pace of algal protein corruption in vitro by proteolytic chemicals, for example, pepsin, pancreatin, and pronase (Fujiwara-Arasaki, 1979; Ryu et al., 1982).

Practically identical outcomes are seen with proteins from green and brown seaweeds. In any case, for these seaweeds, the high phenolic substance may restrain protein accessibility in vivo and, in this way, moderate in vitro figures.

This circumstance is most likely not found for the green and red seaweeds, which have low degrees of phenols and higher protein content.

Among the algal proteins, it is important the occurrence of phycobiliproteins in red and blue-green growth (blue phycocyanin in Spirulin, phycoerythrin in red-green growth) (Boussiba et Richmond, 1979; Fan-Jie et al., 1984). These mixes are comprised of biline (tetrapyrolic open center) connected in a covalant route to a proteic chain.

Ongoing investigations demonstrated that phycobiliproteins present cell reinforcement properties, which could be advantageous in the avoidance or treatment of neuro-degenerative ailments brought about by oxidative pressure (Alzheimer's and Parkinson's) just as in the instances of gastric ulcers and malignancies (Gonzales et al., 1999; Padula et Boiteux, 1999; Remirez et al., 1999).

3.3.4. Lipids and Fatty Acids

Lipids represent just 1–5% of algal dry issue and show a fascinating polyunsaturated unsaturated fat piece especially in regards to with omega 3 and omega 6 acids which assume a job in the avoidance of cardio-vascular infections, osteoarthritis, and diabetes.

The green growth shows intriguing degrees of alpha linolenic corrosive (ω_3 C18:3). The red and brown-green growth are especially wealthy in unsaturated fats with 20 carbon molecules: eicosapentanoïc corrosive (EPA, ω_3 C20:5) and arachidonic corrosive (AA, ω_6 C20:4).

Spirulin gives a fascinating wellspring of gamma linolenic corrosive (GLA) (20 to 25% of the complete lipidic division), which is an antecedent of prostaglandins, leucotriens, and thromboxans engaged with the adjustment of immunological, incendiary, and cardio-vascular reactions. Spirulin along these lines gives an intriguing option in contrast to the next known wellsprings of GLA: onagri oil, blackcurrant pips and borage oil (Renaud et al., 1999; Fleurence et al., 1994).

Other than unsaturated fats, unsaponifiable part of seaweeds was found to contain carotenoids, (for example, B-carotene, lutein, and violaxanthin in red and green seaweeds, fucoxanthin in brown seaweeds), tocopherols, sterols, (for example, fecosterol in brown seaweeds), and terpenoids (Jensen, 1969; Piovetto et Peiffer, 1991; Haugan et Liaaen-Jensen, 1994).

Lipidic concentrates of some eatable seaweeds demonstrated cancer prevention agent movement and a synergistic impact with the tocopherol (Le Tutour, 1990). Seaweeds contain up to 2% of dry load of lipids and a lot of this lipid content is comprised of polyunsaturated unsaturated fats (PUFA). It is presented here that PUFAs represent practically 50% of this lipid content, with quite a bit of it happening as omega-3 and omega-6 lipids. The omega-3 to omega-6 proportion is firmly coordinated, a factor that has been seen as significant in a decent diet.

Both omega-3 and omega-6 unsaturated fats are fundamental, i.e., people must devour them in the eating regimen. Omega-3 and omega-6 vie for the equivalent metabolic chemicals; in this way, the omega-6: omega-3 proportion will altogether impact the proportion of the following eicosanoids. This implies omega-3 and omega-6 ought to be devoured in a fair extent, with the perfect proportion of omega-6: omega-3 going from seaweeds contains numerous basic unsaturated fats, which may add to their viability as a dietary enhancement or as a component of a decent eating regimen.

Seaweeds are additionally regularly tried subsequent to drying, however, the impacts of different kinds of food preparing, for example, canning, have been found to have a negative post-handling impact on unsaturated fat levels.

3.3.5. Seaweeds as Foods

There are various audits accessible on the numerous pharmaceutical and therapeutic bioactive mixes present in seaweeds. This audit centers around the dietary parts of seaweeds as nourishments and their related medical advantages.

Though past surveys have concentrated on the nourishing parts of seaweeds, no evaluations were made of the degrees of supplements contrasted with dietary admission and per-divide sums. Seaweeds are arranged systematically as green growth and they speak to a nutritional category that isn't typically ingested in natural structure to any incredible degree in Western social orders.

Seaweeds are macroalgae and can be isolated into the accompanying four distinctive arrangement gatherings: brown-green growth (phaeophyta), red-green growth (rhodophyta), green growth (chlorophyta), and blue-green growth (cyanophyta).

This audit centers around brown, green, and red-green growth. So as to dodge disarray emerging from the utilization of general terms, for example, 'kelp,' the seaweeds' Latin names are utilized, though basic names are additionally referenced.

The utilization of seaweed in Asian food is notable. In Western social orders, it is essentially connected with restricted beachfront use or authentic stories, yet seaweeds are drawing in expanding consideration as a significant food source.

Common dietary investigations of seaweeds have distinguished significant levels of sugars just as minerals, nutrients, and follow components, for example, iodine. There are two fundamental kinds of polysaccharides in seaweeds auxiliary and capacity.

Auxiliary polysaccharides are similar to earthly plants and are for the most part celluloses, hemicelluloses, and xylans. Capacity polysaccharides, for example, carrageenans, and alginates, are increasingly explicit to seaweed species and speak to the most monetarily abused parts in seaweeds.

These capacity polysaccharides display textural and balancing out properties; hence, they are separated by the hydrocolloid business and are utilized in food applications.

For most populations, fish implies protein as fish, prawns, clams, and other shellfish. Be that as it may, there are numerous other food assets accessible from the ocean. One such model is palatable seaweeds.

This significant asset of the ocean is to a great extent disregarded in run of the mill western weight control plans. By correlation, palatable seaweeds are a basic piece of the eating regimen for individuals who live by the ocean in such territories as Asia, the Pacific Islands including Hawaii, South America, and Africa.

In any case, with the expanding center upon buyer wellbeing and sustenance, seaweeds are currently being reexamined in numerous Western populaces for their nourishing characteristics. Seaweeds are healthfully significant as new or dried, or as a fixing in prepared nourishments.

Eatable seaweeds are awesome wellsprings of nutrients including A, B1, B2, B6, B12, niacin, and C. They are additionally wealthy in iodine, potassium, iron, magnesium, and calcium. Hydroclathrus, Caulerpa, Eucheuma, Gracilaria, and Acanthohora spp. are utilized as green plate of mixed greens vegetables; while coarser Gracilaria and Eucheuma spp. are salted.

3.4. SEAWEEDS AS A SOURCE OF VITAMIN

The living space of seaweeds differs from species to species yet huge numbers of them invest a lot of energy presented to direct daylight in a wet situation. Accordingly, seaweeds contain numerous types of cell reinforcements, including vitamins and defensive colors.

Seaweeds contain both water-and fat-dissolvable nutrients. These incorporate vitamins A, B, C, and E. Vitamin E, a significant cancer prevention agent, is available in an 8 g part of Undaria pinnatifida (Wakame) at a level higher than in peanuts.

Water-dissolvable vitamins, such as, vitamin C are available in huge sums in Ulva lactuca (ocean lettuce), Undaria pinnatifida (Wakame), and Gracilaria spp. at 10 mg per 100 g wet weight, 14 mg for each 100 g wet weight, 15 and 25 mg for each 100 g wet weight, respectively.

In a for every bit correlation, 8 g of Porphyra umbilicalis (Nori) gives 9 mg of vitamin C, or 15% of the RNI.12,15 This adds to the proof that seaweeds can add to healthful admission in a fair eating regimen.

Seaweeds are additionally one of only a handful hardly any vegetable wellsprings of vitamin B12. Ulva lactuca can give this vitamin in overabundance of the suggested dietary recompenses for Ireland of 1.4 g/ day with 5 g in 8 g of dry staple.

This may give a substitute wellspring of vitamin B12 for veggie lovers or vegetarians. The red seaweed of the Gracilaria spp. was likewise as of late found to have 5.4 mg of beta carotene per 100 g DW, which is a moderately significant level contrasted with different vegetables.

3.4.1. Vitamin B12

Algae are a wellspring of nutrients from group B. For example, seaweeds contain vitamin B12, which is especially suggested in the treatment of the impacts of maturing, of interminable fatigue disorder and weakness.

Among the palatable seaweeds, Spirulin is most extravagant in vitamin B12. Every day ingestion of one gram of Spirulin would be sufficient to meet the day by day necessities in B12 (Watanabe et al., 1999).

3.4.2. Vitamin C

Algae give an advantageous source of vitamin C (Qasim et Barkati, 1985). The degrees of Vitamin C normal between 500 to 3000 mg/kg of dry issue for the green and brown algae (a level practically identical with that of parsley, blackcurrant, and peppers), while the red algae contain vitamin C levels of around 100 to 800 mg/kg.

Vitamin C is of enthusiasm for some reasons: it reinforces the resistant safeguard framework, initiates the intestinal ingestion of iron, controls the arrangement of conjunctive tissue and the protidic grid of hard tissue, and furthermore acts in catching free radicals and recovers Vitamin E.

3.4.3. Vitamin E

Because of its cell reinforcement action, vitamin E restrains the oxidation of the low-thickness lipoproteins. It additionally has a significant influence in the arachidonic corrosive chain by restraining the arrangement of prostaglandins and thromboxan.

The brown seaweeds contain more elevated levels of vitamin E than green and red seaweeds. Among the brown algae, the most significant levels are seen in the Fucaceae, Ascophyllum, and Fucus sp., which contain somewhere in the range of 200 and 600 mg of tocopherols/kg of dry issue.

Brown algae contain alpha, beta, and gamma tocopherol while the green and red algae just contain the alpha tocopherol. It was indicated that the gamma and alpha tocopherols increment the creation of nitric oxide (NO) and NO synthase action and furthermore assume a significant job in the avoidance of cardio-vascular infection (Solibami et Kamat, 1985).

3.5. POLYPHENOLS

Algal polyphenols, likewise called phlorotannins, vary from earthly plant polyphenols. Polyphenols from earthbound plants got from gallic and ellagic corrosive, while the algal polyphenols got from phloroglucinol units (1,3,5-trihydroxybenzine).

Phlorotannins establish a very heterogeneous gathering of atoms (structure and polymerization degree heterogeneity) giving a wide scope of potential organic action. Most noteworthy substance are found in brown seaweeds, where phlorotannin extend from 5 to 15% of the dried weight (Ragan et Craigie, 1973; McInnes et al., 1984; Glombitza et Keusgen, 1984).

Cell reinforcement movement of polyphenols separated from brown and red seaweeds has just been exhibited by in vitro examines (Nakamura et al., 1996). Despite that, on the opposite of ground plants polyphenols, little is thought about the job of phlorotannin in forestalling infections connected to oxidative pressure.

3.6. CAROTENOIDS

Carotenoids are incredible cancer prevention agents. Ongoing examinations have indicated the connection between an eating regimen wealthy in carotenoids and a decreasing danger of cardio-vascular malady, tumors (β-carotene, lycopene), just as ophthalmological sicknesses (lutein, zeaxanthin).

Brown seaweeds are especially wealthy in carotenoids particularly in fucoxanthin, β-carotene, violaxanthin. The fundamental carotenoids in the red algae are the β-carotene and α-carotene and their dehydroxylated subsidiaries: zeaxanthin and lutein.

The carotenoid synthesis of the green algae is like that of higher plants: the principle carotenoids present are the β-carotene, lutein, violaxanthin, antheraxanthin, zeaxanthin, and neoxanthin. A great deal of studies showed the cell reinforcement properties of the algal carotenoids and the job they play in forestalling numerous pathologies connected to oxidative pressure (Okuzumi et al., 1993; Yan et al., 1999).

3.7. USE OF VARIOUS SEAWEEDS IN DIETARY SUPPLEMENTS

Seaweed is a staple thing of diet has been utilized in Japan, Korea, and China since ancient occasions. In 600 BC, Sze Teu wrote in China, "Some algae are a delicacy fit for the most respected visitors, in some event, for the King himself."

Some 21 species are utilized in regular cookery in Japan, six of them since the eighth century. Seaweed (Kaiso) represented over 10% of the Japanese eating regimen until moderately as of late, and seaweed utilization arrived at a normal of 3.5 kg per family in 1973, a 20% expansion in 10 years. In spite of the fact that there is little custom of utilizing seaweed in Western cooking, there is presently restored enthusiasm for Western nations in the utilization of seaweed as ocean vegetables.

Lately, there has been a developing enthusiasm for purported useful nutritional categories, among which seaweeds would appear to have the option to assume a significant job since they can give physiological advantages, extra to healthful as, for example, hostile to hypertensive, against oxidant or mitigating (Figure 3.3).

Figure 3.3. Seaweed salad.

Source: Image by Flickr.

A utilitarian food can be characterized as a food that creates a valuable impact in at least one physiological capacity, expands the government assistance and additionally diminishes the danger of experiencing the beginning or improvement of a specific malady. The functionalities are unmistakably more deterrent than remedial.

Besides, new kinds of items, got from food, regularly alluded to as nutraceuticals have as of late been created and advertised widely. These items are generally utilized as food supplements, as opposed to entire nourishments, and are showcased as tablets and pills and can give significant medical advantages (Figure 3.4).

Figure 3.4. Korean seaweed cuisine.

Source: Image by Wikipedia.

Every now and again, useful nourishments are gotten from customary nourishments advanced with a fixing which can give or advance a useful activity for human wellbeing. These are the so-called useful fixings. Concurring Madhusudan et al. numerous organically dynamic mixes are available in seaweed, which can be utilized as restorative operators in dietary enhancements.

3.8. PROTEIN CONTENT

A few seaweeds, for example, Porphyra spp. (Nori) are generally high in proteins. The protein substance can be as high as 47% of the DW, yet these levels change as per the season and the species.

For most species, aspartic, and glutamic acids establish a huge piece of the amino corrosive make-up of these proteins. Aspartic and glutamic acids display fascinating properties with regards to season advancement, and glutamic corrosive is a primary segment in the taste impression of 'Umami.'

Laminaria japonica (Kombu), which is firmly identified with Laminaria digitata, is the first wellspring of the flavor enhancer monosodium glutamate, which was found through its utilization in Asian cooking.

These levels are most noteworthy in brown seaweeds, with red seaweeds having lower measures of these two amino acids. Basic amino acids, for example, histidine, leucine, isoleucine, and valine are available in numerous seaweeds, for example, Palmaria palmata (Dillisk/Dulse) and Ulva spp. (ocean lettuce).

The degrees of isoleucine and threonine in Palmaria palmata are like the levels found in vegetables, and histidine is found in Ulva pertusa at levels like those found in egg proteins. The bioavailability of seaweed proteins can some of the time be hindered by the captured idea of the proteins in the cell network.

Expanding the bioavailability by utilizing physical procedures or maturation to separate the filaments and free more protein has been considered, especially for Palmaria palmata (Dillisk/Dulse), as has the utilization of stomach related enzymes.

The level of absorbability of these proteins is by all accounts identified with the measure of solvent fiber in the algae, forestalling bioavailability of the proteins. Different proteins inside red seaweeds, for example, Chondrus crispus, retain sun-oriented radiation and are commonly known as mycosporine-like amino acids.

There is developing proof that these proteins are strong and stable safeguards of bright light, a trademark that may loan them to use in other areas. Red seaweeds contain a gathering of proteins called biliproteins, some of which are removed for their important use as fluorescent markers.

Phycoethrin is a significant light-gathering shade in red algae and is consistently utilized as a fluorescent test in logical investigations. Different individuals from this gathering of proteins known as phycobiliproteins are phycocyanins and allphycocyanins.

These proteins are collected into a structure in the cell that ingests light where chlorophyll assimilates ineffectively, i.e., the light frequency from

450 nm to 650 nm. This permits some red seaweed to get by in generally profound water, contingent upon obscurity and different conditions.

Seaweed proteins are additionally low in cysteine. Although a couple of seaweeds amass this amino corrosive, most contain low levels. Sulfur, in any case, might be held as glutathione.

Seaweeds might have the option to manage glutathione combination with their high polyphenol levels, which have as of late been connected to expanded articulation of glutamylcysteine. Generally speaking, seaweeds have been investigated well as wellsprings of supplements and proteins for dietary purposes.

3.9. OTHER MAJOR COMPONENTS

Seaweeds contain a lot a larger number of segments than essentially filaments, nutrients, and minerals. The brown seaweeds get their trademark shading from the color fucoxanthin and they contain chlorophylls for photosynthesis.

The more copious segments present in seaweeds incorporate different classes of species-explicit polysaccharides, for example, laminarin in Laminaria species, ulvan in species Ulvales and Floridian starches in red seaweeds.

Alginates involve up to 32% of the dry load of brown seaweeds in Irish waters, and fucoidans are available at levels up to 15% of DW. Laminaria saccharina likewise contains significant sums (up to 14%) of mannitol, which has a sweet taste and gives this species the name 'Sweet Kombu.'

The group of red Porphyra seaweeds contain the polysaccharide 'porphyran' at levels up to 48% by DW, and this polysaccharide has been read for food applications. The event of a starch-based polysaccharide known as Floridian starch or floridoside at up to 42% of dry load in Porphyra spp. also loans this seaweed to use in food.

Palmaria palmata contains up to 45% of the pentose-related polysaccharide xylan. The pervasiveness of xylose and related sugars in this polysaccharide, which is a dietary fiber, may likewise loan this seaweed to advantageous use in food.

Numerous other seaweed-explicit bioactive atoms are available and will be depicted in a further distribution. Palatable seaweeds have been demonstrated to be high in fundamental nutrients and minerals, at levels that would increase a decent eating routine whenever expended consistently.

In particular, follow components and minerals are bottomless in seaweeds contrasted with earthbound staples, and their nonanimal nature loans them to use in numerous food items. The solid tastes related with seaweeds are aligned with the numerous advantageous micronutrients they contain.

Cooking or washing procedures may evacuate huge numbers of these preferences, and the culinary utilization of seaweeds in Asia can be read for instances of this.

Parts of the supplement substance of seaweeds might be increased by the examination accessible on bioactive segments, which have been read for a long time. A mix of these two properties may renew the utilization of seaweed in the recently wellbeing cognizant customer conditions of Western nations.

3.10. BIOCHEMICAL COMPOSITION OF SEAWEEDS

The proximate structure of various types of seaweeds gathered from Mandapam waterfront locales, southeast bank of India was assessed. The greatest protein content was recorded in Phaeophyceae part Padina gymnospora (17.08%) and the base protein fixations were in Chlorophyceaean part Ulva lactuca (3.25%).

In the end the protein substance of Enteromorpha intestinalis (16.38%) (green growth), Gracilaria folifera (6.98%) (Red-green growth) Sargassum tenerimum (12.42%) (Brown-green growth), Codium tomentosum (6.13%) (Green growth), Hypnea valentiae (8.34%) (Red-green growth).

In the end, the high protein content in the brown seaweed Turbinaria luxurious (14.68%) and lower in the red seaweed Gracilaria verrucosa (9.47%) gathered from the Tuticorin shoreline of Southeast India.

Protein content changed among various genera and furthermore in various types of similar variety. This change might be of spatial or fleeting in nature. Notwithstanding, it is to a great extent ascribed to the encompassing water quality as detailed by Dave and Parekh (1975).

The protein substance of green seaweed Ulva rigida gathered from the beachfront area of Chilka Lake of India is 6.64%. So also, the total protein substance of Kappaphycus alvarezii, a red alga gathered from the ocean shoreline of Rameshwaram, Tamil Nadu, India is 18.78%.

Padina tetrastromatica has a place with the request Dictyotales of Phaeophyta is wealthy in amino acids and total eighteen amino acids were accounted for. Notwithstanding lysine, phenylelanine, and aspartic corrosive, histidine also happens as a significant constituent.

The protein substance of red-green growth Gracillari canaliculata from the Massawa District of Red Ocean was recorded as 8.51%. The protein substance of large-scale green growth gathered from the beach of Bay of Bengal in St. Martin's Island was researched.

Among all the species, Sargassum coriifolium has the most noteworthy level of protein (16.07%), while least in Padina tenuis, which was evaluated as 8.32%.

Correspondingly, the total protein in the seaweeds gathered from the Sea of Marmara which is situated in the northwestern Turkey and recorded the most noteworthy total protein esteem among green growth in U. rigida 28.06% and most reduced in U. lactuca 27.7% tested from the surface at Karacaali station.

In the long run, the varieties in total protein esteem contingent upon the profundity of the example gathered. The most elevated total protein among the Rhodophyta is 31.03% in Polysiphonia spp. whereas the least protein is in G. verrucosa (0.94%).

In all Ulva spp. (Chlorophyta) with the exception of U. rigida total protein diminished altogether from 5 to multiple times as indicated by profundity at Ormanli. Be that as it may, in C. tomentosum from Ormanli, there was no profundity subordinate variety.

Chlorophyceae individuals contain high sugar content than Rhodophyceae and Phaeophycean individuals. The most extreme starch content was recorded in the green seaweed E. intestinalis (28.58%) and the brown seaweed Dictyota dichotomy (10.63%) recorded the base an incentive in the green seaweeds U. lactuca (35.27%) and E. intestinalis (30.58%).

The seaweeds gathered from the Maharashtra coast and Kovalam shoreline of India noted most extreme estimation of starch content in Rhodophycean individuals than in Phaeophycean and Chlorophycean individuals.

In the current examination, the contrastingly the high substance of starch in red-green growth may be because of higher phycocolloid content in their cell dividers.

The starch substance of ocean weed gathered from Mandapam coast fluctuated from 20.47 to 23.9%, in that the most extreme sugar fixation was recorded from brown-green growth Turbinaria conoids (23.9%), Sargassum tenerimum (23.55%), Sargassum wightii (23.50%), trailed by the green algae E. intestinalis (23.84%), and red-green growth H. valentiae (23.60%) and Acanthophora spicifera (23.54%).

The base starch content was seen from green alga Codium tomentosum (20.47%) trailed by brown-green growth Padina gymnospora (21.88%), Colpomenia sinuosa (22.46%) and the red alga Gracilaria folifera (22.32%). Correspondingly, the starch substance of green seaweed U. rigida is 22% and in K. alvarezii contains the base starch substance of 2.67%.

The sugar substance of the seaweeds was more extravagant than that of Spirulina. Sugar content was lower (38.94%) in Dictyota dichotoma, however higher (56.29%) in Hypnea musciformis while, in Spirulina sp. it was just 10–20%.

Along these lines, Hypnea sp. was wealthy in sugar when contrasted with different species. Marine full-scale green growth assortments contained low measure of lipids; they are the wellsprings of poly unsaturated fats.

The unsaturated fat dispersion of seaweed items indicated significant level of omega-3 unsaturated fats and showed a healthfully perfect omega-6/omega-3 free unsaturated fat proportion.

The lipid substance of seaweeds changed from 1.33 to 4.6; in that the most extreme lipid content was seen from E. clathrata (4.6%) trailed by G. folifera (3.23%), C. tomentosum (2.53%), C. sinuosa (2.337%) and S. wightii (2.337%).

The base lipid fixation was recorded from E. intestinalis (1.33%) trailed by P. gymnospora (1.4%), S. tenerimum (1.46%) and U. lactuca (1.6%). The total lipid substance of U. rigida is 12% and in K. alvarezii is 1.09%.

The lipid substance of seaweeds gathered from Tuticorin shore of India shifted from 3.15% to 5.30%. The most extreme lipid content was recorded in green seaweed E. intestinalis and the red seaweed G. verrucosa recorded the base substance.

Thus, Chakraborthy, and Santra (2008) have recorded higher lipid content in a similar green seaweeds E. intestinalis (7.13%). Padina tetrastromatica has a place with the request Dictyotales of Phaeophyta is plentiful in water dissolvable nutrients like vitamin B1, B2, B6 and nicotinic acids.

Gas chromatographic investigation was accounted for total sixteen unsaturated fats among which palmatic corrosive is gigantic in sum. Total thirteen mineral components were accounted for among which calcium and Iron were found as significant ones.

The green seaweed U. rigida contains the total free amino corrosive of 8.9%. The debris substance in Hypnea muscifőrmis were lower (11%) and higher (36%) in Padina tenuis.

All the minerals, for example, calcium, magnesium, sodium, potassium, and iron were higher in H. musciformis, H. pannosa, P. tenuis, Sargassum coriifolium and Dictyota dichotoma than that of Spirulina.

In Padina sp., calcium, and magnesium substance per g were high, 48.00, and 44.13 mg, separately. Though, in Spirulina, calcium, and magnesium substance per gram were recorded 4.00 and 4.80 mg, individually. Indeed, the most reduced substance of Ca and Mg in D. dichotoma (4.55 mg) and in H. musciformis (12.64 mg) were higher than that of Spirulina.

Sodium substance was a lot higher (127.65 mg) in H. pannosa than that of Spirulina (7.3 mg) and different species. The potassium substance was 15.2 mg per gram in Spirulina though, H. pannosa and P. tenuis contained 31.91 and 30.00 mg of K. Iron substance were additionally higher in P. tenuis containing 6.64 mg/g while, in Spirulina, it was recorded 1.6 mg/g.

One of the significant employments of the seaweeds is as dietary fiber, as a result of their high substance of polysaccharides. Dietary fiber is a mind-boggling material comprising of the plant cell dividers, basically unpredictable and artificially differing polysaccharides, and other related substances.

The fiber content was higher (8.5%) in Hypnea pannosa followed by Dictyota dichotoma, though it was a lot of lower (2.5%) if there should arise an occurrence of Padina tenuis.

Fucan (polymer of fucose), ulvan (polymer of rhamnose), galactan (polymer of galactose) are the three significant sulfated polysaccharides present in brown-green growth Sargassum sp., green growth Ulva sp., and red-green growth Gelidium sp. separately.

After extraction, total polysacharrides were evaluated in the separate concentrates as 33.24 mg/ml, 25.6 mg/ml, 48.83 mg/ml individually. Presently a-days ocean weeds have been broadly acknowledged by the individuals of beachfront area all through the world because of their significant wellsprings of supplements.

Particularly Padina lenuis and Hypnea spp. are normally eaten entire plants as a decent wellspring of minerals. In Hypneu pannosa, Na substance (127.65 mg/g) were a lot higher, however Ca, Mg, and Fe were higher in Padina tenuis, which was 48.00 mg/g, 44.13 mg/g, and 6.64 mg/g, separately.

Despite the fact that protein substance of these ocean weeds are less, however starch and mineral substance are higher. These ocean weeds might be utilized to take care of the issues of sugar and mineral inadequacy. Red and brown-green growth can amass iodine and, in this way, they might be viewed as a decent wellspring of concentrated iodine.

3.11. NUTRITION'S PROVIDED BY RED SEAWEED

Nori is presumably one of the most notable and popular palatable seaweeds all through Asia and the Western world. The dried sheets are utilized as a "wrap-up vegetable" to cover rice balls containing vegetables in sushi rice, or the Nori may likewise be utilized as food fixings or embellishments. In parts of Europe, nori might be all the more normally known as "purple laver."

While there are numerous types of Porphyra, for Nori the most normally utilized species are Porphyratenera, Porphyra yezoensis and Porphyra umbilicalis. Nori seaweed (Porphyra sp.) has been seen as particularly plentiful in the B complex of nutrients including vitamins B6 and B12.

B12 isn't found in earthbound plants, in western eating regimens vitamin B12 is regularly found in nourishments containing red meat, fish, dairy, and egg items.

Solid degrees of vitamin B12 are related with the recovery of red platelets and a decrease in the potential for frailty. It additionally advances body development and solid working of sensory system.

3.12. NUTRITION'S PROVIDED BY GREEN SEAWEED

These seaweeds incorporate the numerous types of Ulva spp., and are found all through the world including Australia. They are devoured crude or cooked in soups. Ulva spp., are high in protein, solvent dietary fiber and a scope of nutrients and minerals particularly iron (Figure 3.5).

Figure 3.5. Green seaweed dish.

Source: Image by Pikrepo.

Therefore, Ulva sp. has evidently been utilized in multi-nutrients and wellbeing supplements. The comparative in structure Monostroma is generally known as "thin ocean lettuce," and may likewise be usually utilized instead of Ulva.

3.13. NUTRITION'S PROVIDED BY BROWN SEAWEED

Kombu normally refers to huge brown seaweeds or kelps, for example, types of Laminaria. Normal species incorporate Laminaria japonica, Laminaria angustata, and Laminaria saccharina. Most business Kombu or kelp items are sold dried as pieces or powder, and are re-hydrated in water before use.

They are utilized regularly for servings of mixed greens and soups, and are utilized as a sauce. Kombu is a rich wellspring of glutamic corrosive, an amino corrosive liable for "umami" (as it is brought in Japan), or a fifth taste or flavor past the essential tastes of sweet, sharp, salty, and unpleasant.

Wakame (Undaria pinnatifida) is kelp utilized generally in Japan, China, and Korea for soups, as fundamental-dishes and servings of mixed greens. Wakame has an unobtrusive sweet flavor and a tricky surface.

It is eminent for its utilization in soups, for example, miso soup in Japan, and in plates of mixed greens particularly with tofu. A case of this plate of mixed greens is Chuka. Wakame which comprises of destroyed wakame with a sesame seed and stew dressing. This item is brought into numerous western markets including Australia from nations, for example, Japan, and Taiwan. As with Kombu, Wakame is wealthy in polysaccharides and a decent wellspring of dissolvable dietary fiber (Figure 3.6).

Figure 3.6. Brown seaweed as a source of food.

Source: Image by Wikimedia.

Undaria sp. is known additionally to be wealthy in mixes, for example, fucoxanthums, and fucoidans, and these are used as enhancements for against malignant growth and weight reduction.

3.14. CONCLUSION

Presently, seaweeds have been generally acknowledged by the individuals of coastal region all through the world because of their significant sources of supplements and their applications in various work segments.

As far as food, perspective seaweeds may take care of the issues of protein, starch, and mineral inadequacy in human sustenance by devouring them in everyday life. Accordingly, consequences of the current investigation infer that seaweeds are a potential wellbeing food in human eating regimens and might be useful to the food business as a wellspring of fixings with high healthy benefit.

Seaweeds can give a dietary option because of its nourishing and business esteem which upgraded by improving the quality and extending the scope of seaweed-based items.

Despite the fact that there is a lot of the eatable seaweeds yet relatively few investigations had been done on them. So finally, subsequently seaweeds are the "medicines of the future" and for that, various researches are required to be performed on them in order to find their true value.

REFERENCES

1. Kasimala, M. B., Mebrahtu, L., Magoha, P. P., Asgedom, G., & Kasimala, M. B., (2015). A review on biochemical composition and nutritional aspects of seaweeds. Caribbean Journal of Science and Technology, 3, 789–797.
2. MacArtain, P., Gill, C. I., Brooks, M., Campbell, R., & Rowland, I. R., (2007). Nutritional value of edible seaweeds. Nutrition Reviews, 65(12), 535–543.
3. Tiwari, B. K., & Troy, D. J., (2015). Seaweed sustainability-food and nonfood applications. In: Seaweed Sustainability (pp. 1–6). Academic Press.
4. Verkleij, F. N., (1992). Seaweed extracts in agriculture and horticulture: A review. Biological Agriculture and Horticulture, 8(4), 309–324.
5. Wong, K. H., & Cheung, P. C., (2000). Nutritional evaluation of some subtropical red and green seaweeds: Part I-proximate composition, amino acid profiles and some physico-chemical properties. Food Chemistry, 71(4), 475–482.

CHAPTER 4

Food Applications of Seaweed

CONTENTS

This chapter throws light on the use and applications of seaweed in the food industry. This chapter highlights the functions of the seaweeds and its various types of how they contribute to the food applications. It further explains various interventions and how they enhance the human nutrition. Also enhances the importance of the flourishing seaweed as a food source, and gives even the minute details about the seaweeds their antioxidant and nutritive properties. Further, it also explains the various uses of seaweed and highlights the rapid expansion of them and its impact on the health.

4.1. SEAWEED AS A FOOD SEAWEED

Red algae (rhodophyta), green algae (chlorophyta), blue-green algae (cynophyta), and brown algae (phaneophyta) are the groups in which algae is taxonomically classified.

Across the globe, seaweed have a long history of utilizations as a direct or processed food and this macroalgae are also included the above mentioned three groups of seaweed apart from the blue-green algae.

Seaweed in the west is used exclusively used for the extraction of important food hydrocolloids. These hydrocolloids include carrageenan, agar, and alginates. Microalgae are inexpensive products. And the reason behind the inexpensiveness is that is available almost throughout the year and also it can be easily collected and this is the potential of the product.

The microalgae have become a reality and it is done as with the means of advancement of biological and marine sciences, edible microalgae (blue-green algae) are being large-scale cultured. Moreover, in the food application also microalgae has been introduced. Seaweed has been extensively being explored got its merit as a food. It is also considered as a rich source of nutrients. Further, it is also included in the Asian's traditional cuisine. As seaweed is basically regarded as a remarkable marine medicinal food, it has attracted towards the interest of health-conscious societies as it contains biomolecules within itself, apart from its nutritional properties.

The upper layer of the oceans is basically inhabited by algae and further the fact the almost two-thirds of the world is covered by oceans. To obtain photoautotrophic organisms there is an absolute requirement of a large extension. With many advances in this field and various discoveries alga has turned out be a boom in the industry. Algae are considered as one of the valuable products as of economic importance and they are also considered as the high ecological importance (Figure 4.1).

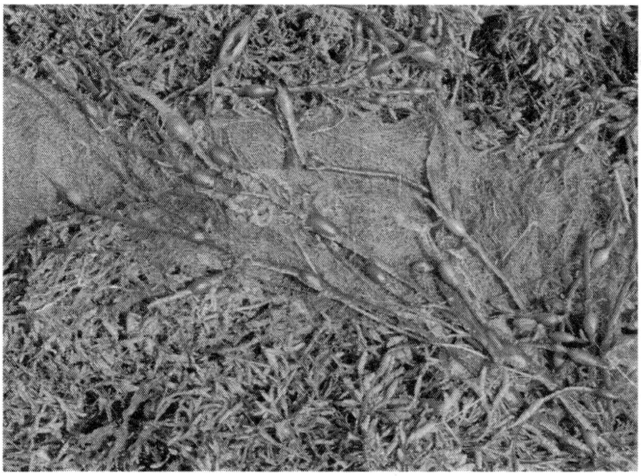

Figure 4.1. Image of seaweed.

Source: Image by Pixabay.

Algae has various functions and these include applications such as in filters or obtaining of minerals, anti-biofilm activity, production of biofuels, fertilizers, as fish feeds, as food or food ingredients, in pharmacology (as they show contraceptive activities or antioxidants).

Algae can help to maintain or even improve human health. New products which are termed as functional foods there is continuous rise in interest. Providing a healthy state is the major application of this functional food.

The wellbeing of consumers can be easily contributed by the characterization of new natural ingredients along with the biological activity and one such example is prebiotic and antioxidant and thus this characterization along with extraction is done in one of the mains areas of the research.

On the chemical composition of macroalgae, the results of a bibliographic survey were reviewed by a present show.

4.1.1. Intervention of Seaweed to Enhance Human Nutrition

The key to wellness of human beings can be considered by the acquirement of a good mental and physical heath through optimum nutrition. A vital role in regulating the bodily functions is done by the nutrients which we obtain from our daily diets.

These nutrients are basically synthesized by using the precursor molecules in our body. Further, these bodily functions are essential for normal development and growth of our body. The human body through different food sources are provided with proteins, carbohydrates, proteins, lipids, and vitamins and also 18 Niranjan Rajapakse and Se-Kwon Kim.

Algae are also considered as a rich source of above nutritional elements, and is an important food source for human nutrition and it can be considered such as when compared with many common vegetables most of the edible seaweed also consists of high levels of fiber, minerals, o-3 fatty acids and also moderate concentrations of proteins and lipids in it.

Further, the factors such as areas of production (Murata and Nakazoe, 2001), variety, and season simply the available amounts of the above nutrients may vary basically. The total carbohydrate content may range from 20% to 76% of dry weight (DW) (Holdt and Kraan, 2011) and also a large amount of carbohydrate as structural, storage, and functional polysaccharides within in the carbohydrates seaweeds depending upon the species.

As it is observed that the polysaccharide dietary fiber, cannot be taken up by the human body, and the greater portion in seaweed is of polysaccharides as the major portion of the seaweed contains high amount of carbohydrate content.

Further, due to this it can be inferred that in terms of bioavailability seaweed is not considered as a good source of carbohydrates. Glucose, mannose, and galactose are comprised in the absorbable forms of carbohydrates in the seaweed.

When compared to the levels of those of FAO/WHO requirements of dietary proteins it is observed that in general seaweed proteins id rich in glycine, arginine, alanine, and glutamic acid and all the essential amino acids are also contained in it (Anonymous, 2006).

Seaweed is observed to be limiting with lysine and cystine when it is compared with the other protein-rich food sources. It was later observed that the essential amino acid index was higher in red seaweed that those in brown and green seaweeds. This observance was made with respect to the protein levels, the amino acid composition, and the amino acid score (Holdt and Kraan, 2011).

Porphyra sp. and Undaria sp., in some red seaweed these were observed in reference to the amino acid score. When compared in quantitative terms to legumes at 30 to 40% of dry matter it is observed that red seaweed basically consists of the highest protein content. Further, in the case of brown and green seaweeds, only 15 to 30% of proteins are observed respectively (Murata and Nakazoe, 2001).

Porphyra palmate is the type if protein content which is available in the red seaweed species and 21% to 47% is the range of Porphyra tenera.

And the range is 7% to 16% observed in case of brown seaweeds, i.e., namely known as Laminaria japonica and Unadaria pinnatifida (Marsham et al., 2007) and for the above-mentioned ranges basically a comparatively study was carried out on the brown seaweeds and red seaweeds as they consist the protein content and help in revealing in these statistics. Further, to include in the diet as a good source of proteins most of the edible red seaweeds could be considered. As compared to the brown seaweed the red seaweed basically consists of less interesting properties in flavor development as these properties are exhibited by the presence of aspartic and glutamic acids and these properties of seaweeds were compared by the nutritional and related health properties.

The high protein content which is nearly close to 70% of dry matter is explicitly present in the blue-green alga, Spirulina, and this high protein content is the reason why these blue-green algae are famous for. It is observed that in the case of seaweed proteins in terms of its vivo digestibility is not well documented. It is also observed that the 80% irrespective of the species are basically classified by the means of the extractability and the in vitro digestibility of seaweed proteins (Fluerence, 1999).

The range from 1% to 5% of dry matter is of the little lipid content on the lipid seaweed (Khotimchenko, 2005). It is also observed that the proportion of essential fatty acids is comparatively high than in land plants, also the major lipid classes in all the seaweeds are basically referred to the neutral lipids and glycolipids.

The total lipid content is elevated during the hot seasons and especially under the cool temperature or climates a higher amount of polyunsaturated fatty acids (PUFAs) are synthesized by the seaweeds (Narayan et al., 2006).

The total fatty acid content nearly contains about 30% of the PUFA as the major of the PUFA in most of the seaweed consists is EPA. The whole seaweeds are balance source of 0–3 and 0–6 acids and also the red seaweeds are also rich in EPA content and 0–6 fatty acids such as arachidonic acid. PUFA when compared to the other foods derived from plant and animal sources and also therefore, seaweed is also a good source of health promoting

PUFA. Approximately 4 to 10% of the total lipid includes the number of phospholipids in seaweed.

The nutritive value of the food is enhanced, and the ease of digestion and absorption of fatty acids is basically done by the phospholipids in the diet and it basically acts as an emulsifier. An efficacy as a part of a balanced diet is mainly added by the means of seaweed as it contains many essential fatty acids.

At varying levels, lipid soluble vitamins such as A and E and also water-soluble B and C are present in the vitamin's seaweeds. The vitamin content (10 mg/100 g) in peanut (Anonymous, 2004) is less than as compared the brown seaweed, U. pinnatifida it contained 14.5 mg/100 g of vitamin E.

To protect PUFA in seaweed is protected by this high vitamin E content and nutritional benefits are maintained by 20 Niranjan Rajapakse and Se-Kwon. The amounts may be range from 20 to 170 ppm and 500 to 3000 ppm, respectively, as the red and brown seaweeds are rich in carotenes (pro-vitamin A) and vitamin C.

In most land plants vitamin B12 are found and they are also considered as a good source, but in few vegetables in the considerable amounts (Bender, 1980). Appreciable amounts of minerals are indicated by the high ash content which is basically contained in the seaweeds generally minerals.

Micronutrients basically include iodine, zinc, copper, selenium, molybdenum, fluoride, manganese, boron, nickel, and cobalt whereas the macronutrients include sodium, calcium, magnesium, potassium, chlorine, sulfur, and phosphorous and further the mineral content of seaweed can account for up to 36% of its dry mass and these mineral micronutrients and macronutrients.

In dry matter among all these, calcium holds up to nearly 4 to 7%. A considerable amount which is compared to the recommended daily allowances (800–1000 mg) (Anonymous, 2004) 7% of calcium, which is considered as a typical daily portion size of seaweed (8 g DW) providing 560 mg of calcium.

The bioavailable form of calcium in milk is calcium carbonate but in the case of seaweeds, the calcium is basically available as calcium phosphate. In some of the seaweeds, the iodine content exceeds its dietary minimum requirements (150 mg/day) but seaweed is considered as the primary source of iodine.

4.1.2. Seaweed as Human Food

Since the prehistoric times, the seaweed, in Japan, Korea, and China has been used in the form of the main item of diet. In the time period of 600 BC, Sze Teu wrote in China, "Some algae are a delicacy fit for most of the valuable guests, even for the King himself." There are near about twenty-one species that are used in cooking in Japan, on a daily basis. Out of which six of them are used since the time period of the 8th century.

Seaweed has been accounted for over 10% of the Japanese diet until relatively recently. In addition to this, the consumption of seaweed has reached an average of 3.5 kg per household in the year of 1973. Which is a 20% increment in the time period of the last one decade (Indergaard, 1983).

Talking about Japan, the most vital or most necessary food species are Nori (Porphyra spp.), Kombu (Laminaria and Saccharina spp.), and wakame (Undaria pinnatifida).

With respect to the West, seaweed is mainly considered as a healthy food item, and also, there has been an increase in the interest in the seaweed as food in the time span of last two decades. It is not likely to be that, the consumption of the seaweed in west will ever be more than a fraction when it is compared with the consumption rate in Japan.

Rules and policies for the collection of Palmaria palmata (dulse, dillisk), a red seaweed, are stated in the Icelandic stories of the 10th century. The seaweed, which is edible in nature, has been also used in Scotland along with Ireland, and the consumption of seaweed is remains for even longer interval of time.

Moreover, in the present interval of time, Palmaria palmata (dulse) is becoming more and more common again in Ireland as well as in the United States.

Laver (Porphyra) has been consumed in Wales from the time period of at least A.D. 1600. Chondrus crispus (Irish moss) was suggested as a healthy food item in Ireland at the starting of the 19th century (Mitchell and Guiry, 1983) (Figure 4.2).

Figure 4.2. Seaweed as human food.

Source: Image by Wikimedia Commons.

Several kinds of various red algae have been used in the Mediterranean in the form of sources of dying agents as well as in the form of the other health-related items, from the time period of pre-Christian (Stein and Borden, 1984).

The usage of the kelps (in Japan, it is known as Kombu, and in China, it is known as kunbu or haidai) dates back to at least the 5th century in China (Tseng, 1981a, b). The species that were used in the form of prime ingredient is Saccharina japonica (Laminariales), in addition to this, there are near about eight to eleven other species are also used, primarily in Japan.

Plants are supposed to dry after the process of harvesting and either chopped into the form of strips or in the form of powder. With respect to Japan, the kombu is used in the making of dishes related to fish, meat dishes, along with a vegetable and rice as well. On the other hand, the Kombu, in the form of powder is used wither in sauces and soups or it is added to rice, just in a similar manner as curry.

Kombu can also be used through the course of cooking beans in order to add some nutrients and enhance the digestive capability of the dish. Moreover, Kombu is also a rich source of glutamic acid; it is used in order to enhance the flavor, as well as it is a natural alternative of monosodium glutamate. Some types are also employed in order to make an infusion which is identical to tea.

With respect to the Korean cuisines and Chinese cuisines, Kombu plays an important role. In the time period of 1976, near about 176,000 wet tonnes of Laminaria spp. were gathered from the wild sources in Japan, out of which, near about 22,000 tonnes of it were cultivated.

Laminaria spp. is cultivated either with the help of the seeding blasted areas of rocky shores or with the help of seeding ropes. Talking about China, Saccharina japonica (Laminaria japonica) was imported from Japan from the 5th century until the establishment of the People's Republic. In the time period of early 1950s, people of China started to cultivate this species, which had been accidentally introduced from Japan to Dalian on the Yellow Sea. In the present interval of time, near about 6 million tonnes wet weight of this kelp are cultivated in China. The strains grow may depict more than a single species.

The other form of kelp, Undaria pinnatifida (which is also a Laminariales), is extensively used in Japan. This species is a more significant crop when it is compared with Laminaria spp., with respect to both values as well as production (Tseng, 1981b).

The increment in the natural production were accomplished for several numbers of years with the help of placing stones on the bottom of the sea and blasting rocky reefs in order to elevate the area which is appropriate for the attachment. Artificial seeding is carried out on cleared areas with the help of either zoospore suspensions or with the help of sporophylls (which is a specialized leaflet that bears the zoosporangia).

In the time period of 1960 to the year 1969, the annual production from the natural habitats was near about forty thousand to sixty thousand wet tonnes. Rope cultivation has been carried out from the time period of 1955, and the ropes were seeded with the help of attaching sporophylls. Fusions with greater growth as well as nutritional values have been developed in Japan.

In the year of 1976, near about twenty thousand wet tonnes were collected from the wild source and near about 127,000 wet tonnes were cultivated. The algae are harvested and dried after washing in freshwater. After re-soaking the plant material is used in the form of an additives into the soups (wakame soup), toasted (Yaki-wakame); used half re-soaked, with boiled rice; and coated in sugar and tinned (Ito-wakame).

With respect to China, Undaria pinnatifida was gathered with the help of natural habitats for centuries, and primarily on the East China Sea coast. Plants are grown now on ropes in the Quingdao and Dalian areas (Yellow Sea), to where the algae were transplanted from Korea (Tseng, 1982).

Undaria sp. is not as common as Laminaria sp. is common in China, in the form of a foodstuff and the growers find it challenging tasks in order to manage the plants. In this way, in China, the annual production is very low, that accounts to no more than a few hundred tonnes in DW which each passing year.

Generally, Wakame is used in soups, for example, miso soup or in sunomono salads. According to the studies that were directed in the Hokkaido University have found that the fucoxanthin present in wakame can be used in order to burn fatty tissues. Moreover, according to the studies which were conducted on mice basically depicted that the fucoxanthin makes expression of the fat-burning protein UCP1 (uncoupling protein). That accumulates in fat tissue around the internal organs. Expressions of UCP1 were elevated in the considerable amount in mice that have fed with fucoxanthin.

Laverbread (Porphyra dioica) is picked on the coast of Wales from intertidal rocks and is known in Welsh as "bara lawr." Laverbread or bara lawr has an excellent nutritional content, primarily protein, some kinds of carbohydrate, vitamins A, B, B2 and C along with lots of trace elements and minerals, out of which, the most prominent is iodine.

Apart from this, the other plus point is that, Laverbread or bara lawr is low in calories and very apt for all sorts of vegetarians. In Wales, it is finding in the form of tablet or tinned in any kind of health shop. When the seaweed arrives at the Welsh factories, it is examined and goes through a series of washes. It is then cooked for couple of hours with only a little salt added. As an outcome, this produces a very dark, dense spinach-like purée.

Several numbers of chefs are exploring the culinary properties of the laverbread, and adding new dimensions to their lists. Conventionally, laverbread has been served as a very simply spread on thin toast with oatmeal or malted vinegar, and served with fried or grilled Welsh-cured bacon. In the latest times, laverbread has been served with pasta dishes as well, in spicy batter accompanied with mushrooms, as well as seafood pizzas.

There is now renewed interest in the Western nations, with respect to the use of seaweeds in the form of vegetables.

4.1.3. Antioxidant Activities of Seaweed

Antioxidants are very attractive when they are present in the form of supplement. And the reason behind this is their potential preventive role in various kinds of diseases which are related to the oxidative stress, it takes place when the balance between the antioxidants and the reactive oxygen species (ROS) is disturbed due to either, the depletion of the antioxidants or the accumulation of the ROS.

There are several numbers in vitro studies that permit the speculation in this sense, never the less, the clinical or biological proofs for some these sorts of molecules are weak. For example, in the year 2007, according to a systematic review, that states the impact of the supplements of the antioxidant, for example, β carotene, vitamins A, E, and C, and selenium, on overall mortality in primary or secondary prevention randomized clinical trials. It resolved not only that β carotene and vitamins A and E do not have beneficial impacts with respect to the mortality, but that they seem to increase the risk of death.

Certainly, most of the trials that are consisting in this review found out the impacts of the supplements administrated in higher doses as compared to those that are generally found in a balanced diet, and some of the trials used doses well above the recommended daily allowances and even above the upper tolerable intake levels. In addition to this, very heterogeneous populations have been assessed in this review and the effect of several numbers of various kinds of supplements was evaluated in the general population or in patients with gastrointestinal, cardiovascular, neurological, skin, ocular, renal, endocrinological, rheumatoid, and undefined diseases in a stable phase.

Never the less, one likely efficient approach for or treating these kinds of disorders that are mediated with the help of ROS is based on a diet which is full of natural antioxidants, which is aided with the help of several numbers of various international health agencies.

With respect to this context, there is an increment in the interest in the natural antioxidants as the substitute of the synthetic compounds, and the reason behind this is the increased safety concerns and a trend in the direction of the use of natural additives in foods, all across the globe. In addition to this, natural antioxidants are derived with the help of several numbers of different plants and marine algae not only have shown the health-promoting benefits, but they have also demonstrated a great potential for enhancing the oxidative stability with respect to the food products.

In the extreme settings, several numbers of various types of seaweeds that are edible in nature, which is consisting of Fucus vesiculosus, develop distinct metabolic system in order to survive, leading to synthesis of a high number of secondary metabolites, most of which are potent antioxidant molecules.

Seaweeds are known as a rich source of nutrients and of various types of bioactive substances, which is consisting of sulfated polysaccharides, for example, fucoidans, carotenoid pigments, for example, fucoxanthin, and phlorotannins, a subgroup of polyphloroglucinol polyphenols that is only present in the brown seaweeds, that is having potential health benefits.

The antioxidant activity of phlorotannins is closed to phenol rings, which play the role of electron traps to scavenge ROS. Phlorotannins have been found in order to possess several numbers of various physiological activities, with anti-carcinogenic, antibacterial, antiviral, anticancer, and anti-inflammatory characteristics.

In addition to this, polysaccharides, for example, fucoidans, are particularly abundant in seaweed, more particularly in the F. vesiculosus, and also they may act in the form of antioxidants with the help of either directly scavenging ROS, or with the help of induction of the activity of cellular endogenous antioxidant defenses, that are consisting of superoxide dismutase (SOD), catalase (CAT), glutathione transferase, along with the glucose-6-phosphate dehydrogenase.

Seaweeds are also a good source of dietary fibers, prebiotics, as well as other functional ingredients that bring a reduction of glucose as well as reduction of cholesterol in the blood levels. Certainly, the consumption of seaweed has been associated to a lower incidence of chronic diseases, for example, dyslipidemia, and coronary heart disease.

The consumption as well as the commercial use of seaweeds is abundant in Eastern tradition. With respect to these Western countries, seaweeds are specifically employed for the production of phycocolloids. And the reason behind this is their excellent gel characteristics, the polysaccharide fibers of seaweeds, more particularly alginic acid, have also been employed in the form of stabilizing and water-holding agents.

Due to this reason, seaweeds play a very vital role as industrial components in several numbers of various sectors, which is consisting of cosmetic and pharmaceutical/medical, in addition to this, seaweeds are also used in food industry in the form of thickeners, gels, emulsifiers, as well as stabilizers.

Never the less, as the biological activities of seaweeds aid for their potential role in the form of a natural antioxidant, seaweed extracts, and purified compounds can be employed in the form of active ingredients in order to enhance the oxidative stability of functional foods as well as nutraceuticals.

From the time period of 1997, according to the scientific as well as technological developments, the Regulation (EU) 2015/2283 of the European Parliament and of the Council reviews, make clear and updates the classes of food that constitute novel foods. For this Regulation, food that are having the content of fungi or algae, isolated or produced out of microorganisms, and they are defined as "novel food."

Then, never the less, the scientific research provide highlights several kinds of various bioactivity in seaweed species, marketing it as a novel or functional food with health claims necessitates the scientific proof, which must be given with the help of an application that is submitted to EU, this is an extensive as well as a time-consuming procedure.

With respect to this contest, some of the F. vesiculosus extract or purified molecule is already known as generally recognized as safe (GRAS) or novel foods. Baked foods, for example, snacks, cookies, along with the biscuits, that are eaten and stored for long interval of time before consumption, are require to preserve their quality in order to maintain the competitive aspect in the economic market.

In order to have better shelf life, the use of antioxidants, antimicrobials, as well as anti-browning additives are done by the industry of food. The use of the synthetic antioxidants has been associated to the possible toxicity and side effects, for example, carcinogenesis, and the application of the synthetic antioxidants has reduced. And the reason behind this is the consumer awareness as well as demand for natural protection.

Only fewer natural food antioxidants are commercially available on the market. Out of these, rosemary extracts have been the most successful natural plant-based antioxidants commercialized.

F. vesiculosus is a brown algae species which have high antioxidant activity, apart from the other distinctive characteristics (for instance, anti-inflammatory, and anti-diabetic activities). These properties make it specifically more attractive for its Antioxidants, in which three out of fifteen are employed in several food systems.

Because of the strong market demand as well as, very positive preliminary tests, it is believed that its extracts can be highly competitive on the market and find several numbers of uses in food. In this topic of the chapter, the bioactive properties of F. vesiculosus have been described. In addition to this, the extracts of the F. vesiculosus with the help of several numbers of various chemical as well cell-based methods in order to provide proofs of possible antioxidant mechanism(s) has been described.

4.2. ALGAE AQUACULTURE

Aquaculture is the field of science that is related to the cultivation of aquatic animals, plants, and correlated organisms such as fish, shellfish, seaweed, as well as microalgae. With respect to the human use as well as the consumption, aquaculture would be a fast-growing industry.

In the present interval of time, several numbers of different authors are concentrated on performing life cycle assessment (LCA) of seafood production in order to provide new visions and understandings into its environmental effects and in this way, in order to enhance the environmental sustainability of the aquaculture production systems.

Never the less, the technologies that are associated to the aquaculture of algae have experienced considerable amount of development in the time period of past seven decades, more particularly in the several regions of Asia. Moreover, also in America and Europe, there is still much to enhance with the respect to their science as well as the social acceptance that requires.

Out of many, one of the main difficulties is the development of strains that are thermo-resistant, of quick growth, with high production of compounds of interest, resistance to the morbidity, and antifouling capacity, along with the development of effective as well as economical hatcheries, that are capable to survive the storms in the open sea.

There is one fact in order to keep in mind that, the algae aquaculture provides several benefits to the ecosystems, as it enhances the situations of the coastal waters, favoring the other species as well as the environment. Fascinatingly, even though over ten thousand species of algae are known, aquaculture of algae is mostly made (around 81% of the production) with the help of very few of them:

The brown algae are Saccharina japonica, Undaria pinnatifida, red algae Porphyra sp., Kappaphycus alvarezii and Eucheuma striatum (car rageenophytes), and Gracilaria/Gracilariopsis spp. (agarophytes).

Out of many, one of the primacies when developing aquaculture is to be sustainable, that is, in order to make sure the minimum possible adverse impacts on the environment. One technique in order to accomplish this is through the development of enhanced methodologies or tactics for waste treatment.

Integrated multi-trophic aquaculture (IMTA) is one interesting way to develop an ecosystem more balanced. This system integrates the production of fish with seaweeds, so the seaweeds use the nutrient waste fish feed to grow (Figure 4.3).

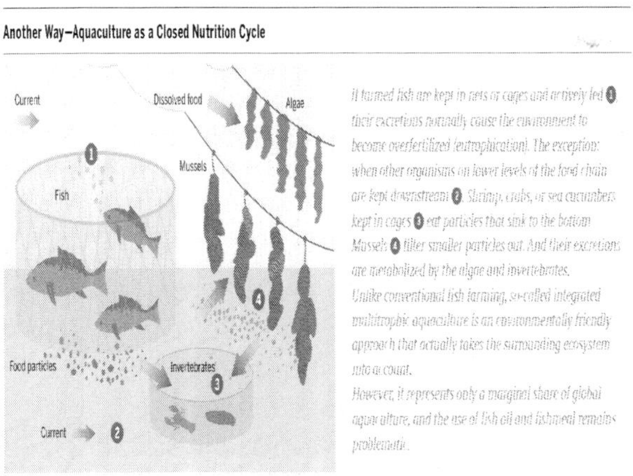

Figure 4.3. Algae aquaculture.

Source: Image by Wikimedia Commons.

4.2.1. Nutritional Composition of Edible Seaweeds

Proximate composition (moisture, ash, protein, and oil content), total dietary content of fiber and physico-chemical properties of three brown and two red edible Spanish seaweeds, namely: Himanthalia elongata (sea spaghetti), Bifurcaria bifurcata, Laminaria saccharina (sweet kombu), Mastocarpus stellatus and Gigartina pistillata were studied. Ashes (near about 24.9% to 36.4%) were high in all samples.

The content of the protein ranges from near about 10.9 to 25.7%, which is much higher with respect to the Laminaria (25.7%). Further, followed with the help of red seaweeds which are basically about 15.5 to 21.3%. Minor components were lipids (near about 0.3 to 0.9%) in all samples, apart from Bifurcaria sp. (near about 5.6%).

There are a little over 7 grams of carbohydrate in 1 cup of raw seaweed. Of this, less than 1 gram comes from fiber and sugar combined. Seaweed contains various polysaccharides that act as antioxidants, providing numerous health benefits.

In the conclusion, these sorts of seaweeds can be approximated in the form of good source of food fiber, protein, and minerals for human consumption (Gómez-Ordóñez et al., 2010). The content of mineral that is present in them was identified in various brown (Fucus vesiculosus, Laminaria digitata, Undaria pinnatifida) and red (Chondrus crispus, Porphyra tenera) marine sea vegetables that are edible in nature.

Seaweeds are consisting of high amounts of ash (which ranges from near about 21.1 to 39.3%) and sulfate (that is ranges near about 1.3 to 5.9%). With respect to the brown algae, the content of ash present in them ranges from 30.1%, which is higher when it is compared with the red algae, that ranges from 20.6% to 21.1%.

Edible brown and red seaweeds can be consumed in the form of a food supplement in order to provide assistance to fulfill the required intake of some essential minerals and trace elements on daily basis (Rupérez, 2002). Sea spaghetti (Himanthalia elongata), Wakame (Undaria pinnatifida), and Nori (Porphyra umbilicalis), on fatty acid composition, amino acid profile, protein score, mineral content, and antioxidant capacity in low-salt meat emulsion model systems.

A decrease in the w-6/w-3 PUFA ratio and an increase in w-3 polyun saturated fatty acids are observed by the addition of seaweeds. Further increase in concentrations of K, Ca, Mg, and Mn is observed by the addition of 740 food industries which leads to seaweed productions.

No significant changes are observed in the amino acid profiles in the model systems by Wakame and Sea Spaghetti whereas on the other hand an increase in the levels of serine, glycine, alanine, valine, tyrosine, arginnine, and phenylalanine is observed due to the presence of Nori (Lopez et. al, 2009).

Further, Laminaria sp., Undaria pinnatifida, Sargassum fusiforme and Porphyra sp. together basically produce 34 edible seaweed products and the nutritional compositions of these products were already analyzed.

When brown algae varieties were compared to the red algae it was observed that these consists of the all the essential amino acids which were basically detected and further uniquely high concentrations of taurine was also featured in this red alga (Dawczynski et al., 2007).

H. elongata, Laminaria ochroleuca, Undaria pinnatifida, Palmaria sp., Porphyra sp., Saccorhiza polyschides and Himanthalia elongata are some of the total lipid, protein, ash, and individual fatty acid contents of edible seaweeds and these can be easily canned and also supplied conveniently.

The omega 3 and omega 6 acids are the balanced sources of the red seaweeds and the brown seaweeds. Saturated fatty acids are present in the red seaweeds and in the brown seaweeds; unsaturated fatty acids are predominantly present in it.

4.3. USES OF SEAWEEDS

In the sixth century in China and in fourth century in Japan, the use of seaweed as food has been tracked by them. The largest consumers of seaweeds in the world today are these two countries Japan, China, and along with them is the Republic of Korea.

The demand of seaweeds has been followed to some other parts of the world by the nationals from these countries that basically migrated to other parts of the worlds. Further, there are few examples such as countries like South America and the United States of America.

The ability to supply requirements from natural (wild) stocks has basically outstripped the increasing demands over the last fifty years. The cultivation industries that now produce more than 90% of the market's demand were basically developed by the means of the research into the life cycles of those seaweeds.

A different type of market was basically developed in the countries such as Ireland and Iceland and the reason behind this is that entirely a different type of seaweed has been traditionally by the people living these countries.

Promotion of seaweeds for the restraints and domestic use with some success was mainly done by some of the governmental and commercial organizations in France. In some areas, there has been a tradition of using fresh seaweeds as vegetables and in salads and for this, basically, an informal market exists among the coastal dwellers in some of the developing countries.

Almost 5 million wet tones of edible seaweeds have been harvested by the largest producer of it and this country is China. Introduction to commercial seaweeds was basically provided by the means of a guide to the seaweed industry.

A species of Porphyra is red seaweed and nori is produced by this red seaweed. Nori is basically the thin dark seaweed which is wrapped around a rice bell in sushi. 75% of Nori and 600,000 wet tones is Japanese production.

Mainly three types of hydrocolloids which are namely are alginate; agar and carrageenan were produced by the various red and brown seeds.

Hydrocolloids dissolve in water to give a thickened (viscous) solution. These are basically noncrystalline substance with very large molecules. To form gels (jellies) of varying degrees of firmness, and also to form water-soluble films and some products such as ice creams (they inhibit the formation of large ice crystals so that the ice cream can retain smooth texture) are also stabilized by the Alginate, agar, and carrageen. These three are basically water-soluble carbohydrates.

In Japan firstly discovered that the gelling properties of agar can be easily extracted with the help of hot water from red seaweed. In the date back to 1658, seaweeds were basically considered as the source of hydrocolloids. In the nineteenth century, the most popular thickening agents were the extracts of Irish moss, red seaweed which basically contains carrageenan. Further, the thickening and the gelling agents were not produced commercially and sold until 1930 and these were basically the extracts of brown seaweeds which contain alginate. After the Second World War, it was observed that the industrial use of the seaweed's extraction expanded rapidly, but by the availability of the raw materials, it was sometimes limited.

Cultivation industries basically now day's supplies high proportion of the raw materials for the hydrocolloids. Also, these cultivation industries were developed by the research done once again on the life cycles.

The above three hydrocolloids were produced today by the extraction and harvest of approximately 1 million tons of wet seaweeds. With a value of nearly US$ 585 million, the total hydrocolloid which is produced is nearly 55000 tonnes. When the brown seeds are extracted then alignate production is done by them (US$ 213 million).

These brown seeds are mainly harvested from the wild. Moreover, for the industrial uses the cultivation of brown seeds for the raw material can be expensive.

The expansion of the agar industry is only due to the cultivation of agar and this cultivation is present only since 1990. Further, only two types of red seaweeds are mainly produced by the agar (US$ 132 million). And out of the types of the red seaweed only one of them are cultivated since the 1960–1970.

Irish moss is basically small seaweed which is grown in cold water along with a limited resource base. Moreover, especially on this Irish moss, the carrageenan production depends (US$ 240 million) and this production somewhat also depends on the wild seaweeds. The carrageenan seaweeds were successfully been cultivated in warm water countries along with low labor costs. Also, the availability of these carrageenan seaweeds was the reason for the rapid expansion of the industry since the early 1970s.

Although there is an explicit demand for some of the wild species from South America and Irish moss, in the current scenario the most of the seaweeds are used for the production carrageenan with the means of cultivation.

In the 1960s, the production of seaweed was pioneered in Norway, also these seaweed meals were used as an additive to the animal feed. Brown seaweeds are used to make it and these brown seaweeds are firstly collected, dried, and milled. The crude oil prices are the reasons which usually affect the cost and also oil-fired furnaces are usually used for the process of drying of these brown seaweeds.

Usually, large brown seaweeds were collected and dug into the local soil and this was basically by the early usage done by the coastal dwellers and these collected brown seaweeds were basically the storm cast seaweed.

While the mineral content is a useful fertilizer and source of trace elements, and on the other hand, the high fiber content of the seaweeds acts as a soil conditione,r and the moisture retention is assisted. A small industry was dwindled with advent of 2A guide to the seaweed industry which was the introduction to the storm cast commercial seaweeds 3 synthetic chemical fertilizers and this small industry was basically developed in the early twentieth century and it was based on the drying and milling of mainly storm cast material.

The combined state combined costs of drying and transportation have confined usage to sunnier climates where the buyers are not too distant from the coast. Not on the large scale but on the other hand in today's world the rising popularity of organic farming is leading to some revival of the industry.

In the production of liquid seaweed extracts basically the growth area in seaweed fertilizers in this. For the dilution by the user, these can be produced in concentrated form. The plants can be watered in around the roots, and several can also be applied directly on to the plants.

These products can be effective and this was mainly be proved by the several scientific studies. For 10,00 tonnes of seaweed extracts which values nearly US$ 5 million an estimation was made that nearly about 10,000 tone of wet seaweed were used and it was made in 1991.

Because of the wider recognition of the usefulness of and organic farming was rapidly increasing popularity the market has probably doubled within the last decades. This recognition was made especially effective in the growing of vegetables and some fruits. Marine extract and extract of alga, seaweed extract, or similar are the types of the labels which are included and these labels are of the cosmetic products such as creams and the lotions. From the seaweed, one of the hydrocolloids is extracted from them and this is mainly inferred by the above-mentioned text. The skin moisture retention properties of the products could be improved by the means of Alginate or carrageenan. In thalassotherapy, the pastes of seaweeds are used and these are mainly produced by the freeze crushing or by the cold grinding. This paste off seaweed is basically used for applying on the person's body and then infrared radiation is used for warming.

For a relief in case of the rheumatism and osteoporosis, it is further said that this treatment in conjunction with the seawater hydrotherapy could be used. To generate methane gas for use as a fuel the idea was to grow large quantities of seaweeds in the ocean and ferment (Figure 4.4).

Figure 4.4. Seaweed as fertilizers.

Sources: Image by Wikimedia Commons.

4.4. HEALTH IMPACT OF HABITUAL INTAKE OF WHOLE SEAWEED

The potential risks and benefits of consuming of seaweed components, some observational evidence does exist and these are observed while reviewing the highlights of the lack of human intervention trials. All-cause mortality and cardiovascular disease mortality in Japanese are basically associated with the means of epidemiological which basically indicates that seaweed containing diets which link to them.

Although it has been observed that no mechanistic insights have been published but on it has been noted in the case of Korean men that the one without metabolic syndrome consumes less seaweed as compared to the other hand Korean with metabolic syndrome are reported to significantly consume more seaweed. No association was founded between Undaria pinnatifida.

This was basically demonstrated by the University college cork user in the year 2019 march 19. Further, it was observed that in the case of premenopausal women along with the risk of breast cancer was basically inversely associated with consumption of Porphyra. And this relation was absent in the case of the postmenopausal women. Moreover, breast cancer risks and the Undaria pinnatifida were not at all associated with each other.

In the case of postmenopausal women, a study done by Michikawa et al. the risk of thyroid cancer and between the consumption of the seaweed and in these women the cancer is especially papillary carcinoma. Colon cancer and stomach and also between the seaweed consumption there was an inverse relationship was implicated by the means of case-controlled studies which was done by Hoshiyama et al. (2006).

On the other hand, it was observed in the low sample of the studies there was a sense of vigilance depicted by the means of interpretation. In pregnant Japanese women (n ¼ 1002) dominance was observed regarding the allergic rhinitis, further also between Sargassum fusiforme, Undaria pinnatifida and Porphyra sp. an inverse association was also reported.

Since there are currently recommendations regarding brown seaweed intake in the case of pregnant women in countries namely Australia and New Zealand. Further, the studies which are mentioned above did not contribute to the knowledge of iodine intake from seaweeds during pregnancy as these studies did not measure the iodine intake or iodine status.

The reports of non-virus contamination of Entermorpha spp. depicted some of the concerns regarding the potential for the seaweed which could further contribute to the foodborne infections, further in a toxin in also Gracilaria edulis and the presence of polycavernoside.

As serum TSH could significantly increase (within the normal clinical range), when it followed by the intake of palmaria palmate 5 g/d for 28 days due to this a recent study on intervention study in a European population basically concluded that the iodine status in adults could improve by the consumption of that Palmaria palmate.

The seaweed could affect health, immune function, and disease risk and this was component by undertaking the human interventions to establish these factors, and these undertakings were basically highlighted by the authors which need to characterize seaweed consumption.

4.5. CONCLUSION

Excessive intakes of salt, iodine, and heavy metals should be safeguarded. And this could be done by the ensuring the various legislative measures such as monitoring and labeling of food products. The seaweeds basically contribute to the future global food security is was the reason to ensure the legislative measures.

Mainly edible seaweeds are a rich and sustainable source of macronutrients (particularly dietary fibers) and micronutrients. More studies of heavy metal toxic kinetics are required, further bioaccumulation of arsenic is a risk due to the heavy metal concentrations in the edible seaweeds which are generally observed below the toxic levels.

The recommendation portion size could be limited by the means of obtaining the amounts of PUFAs, protein, or dietary fibers, and these are basically obtained by the means of a tradeoff between iodine and heavy metals ingestion. Also, to obtain meaningful amounts of these components the whole seaweeds is needed. Further, there are various studies which have been increased particularly regarding the potential agents to combat noncommunicable diseases; also, it is increasing as the novel bioactive components are extracted from seaweeds.

The added value products are mainly created by the extraction of individual components from the complex seaweed matrix and this is a justifiable strategy.

Nowadays, on the basis of many industrial applications the seaweeds are being studied. These industrial products are namely cosmetics, food, chemistry, paint, medicine, etc. By using the compounds such as pharmaceutical, cosmetics, and food industries the extraction of compounds are done within the western countries (Wijesinghe and Jeon, 2012a).

In the degenerative diseases such as cancer, inflammation, arthritis, diabetes, and hypertension a significant role in prevention of these diseases by the biological active compounds of seaweeds and these compounds are namely carotenoids, alginic acid, fucoidan, peptides, and phlorotannins.

For the functional ingredients in many industrial applications which are namely cosmeceutical, pharmaceuticals, and functional foods and for this seaweed serve as a rich source of novel chemical entities.

These seaweeds are mainly derived from the active components and these are also immense biochemical diversity (Wijesinghe and Jeon, 2012b). Further, the commonly used varieties which are available commercially and used as seaweeds are of marine macroalgae.

The two main algae which are used as the human food sources these are namely red algae and brown algae. Factors such as season and environmental growth conditions are the main factors on which the protein content of seaweed species varies and greatly and these protein content also depends on these factors.

REFERENCES

1. (2018). Eco. Env. & Cons., A Sustainable Resource for Food and Pharmaceutical Industry in India, 24(1), pp. 246–254. Copyright@ EM Internationalism 0971–765X, corresponding author's E-Mail: Goraadnan@Gmail.Comseaweeds. [online] Available at: https://www.researchgate.net/publication/323143450_SEAWEEDS_A_SUSTAINABLE_RESOURCE_FOR_FOOD_AND_PHARMACEUTICAL_INDUSTRY_IN_INDIA (accessed on 10 July 2020).
2. Gomez-Zavaglia, A., Prieto Lage, M., Jimenez-Lopez, C., Mejuto, J. and Simal-Gandara, J., 2019. The Potential of Seaweeds as a Source of Functional Ingredients of Prebiotic and Antioxidant Value. Antioxidants, [online] 8(9), p.406. Available at: <https://www.researchgate.net/publication/335947593_The_Potential_of_Seaweeds_as_a_Source_of_Functional_Ingredients_of_Prebiotic_and_Antioxidant_Value> [Accessed 16 July 2020].
3. (n.d.). A Guide to the Seaweed Industry. [eBook] p. 4. Available at: http://www.fao.org/3/a-y4765e.pdf (accessed on 10 July 2020).
4. Corsetto, P., Montorfano, G., Zava, S., Colombo, I., Ingadottir, B., Jonsdottir, R., Sveinsdottir, K. and Rizzo, A., 2020. Characterization of Antioxidant Potential of Seaweed Extracts for Enrichment of Convenience Food. Antioxidants, [online] 9(3), p.249. Available at: <https://www.ncbi.nlm.nih.gov/pmc/articles/PMC7139466/> [Accessed 16 July 2020].
5. (n.d.). Nutritional and Digestive Health Benefits of Seaweed. [eBook] p. 3. Available at: http://marineagronomy.org/sites/default/files/Digestive%20Health%20Seaweed.pdf (accessed on 10 July 2020).
6. Cherry, P., O'Hara, C., Magee, P., McSorley, E. and Allsopp, P., 2019. Risks and benefits of consuming edible seaweeds. Nutrition Reviews, [online] 77(5), pp.307-329. Available at: <https://pubmed.ncbi.nlm.nih.gov/30840077/> [Accessed 16 July 2020].
7. (n.d.). Seaweeds for Food and Industrial Applications. [eBook] p. 3. Available at: https://cdn.intechopen.com/pdfs/41694/InTech-Seaweeds_for_food_and_industrial_applications.pdf (accessed on 10 July 2020).
8. Seaweed.ie. (n.d.). Seaweed.ie: Seaweed as Human Food. [online] Available at: http://www.seaweed.ie/uses_general/humanfood.php (accessed on 10 July 2020).

CHAPTER 5

Seaweeds in Agricultural Use

CONTENTS

In this chapter, the various uses of seaweed in agriculture has been discuses, in addition to the impact of seaweed on agricultural crop plants as a biofertilizer.

In this chapter, the effect of seaweed on diseases and insects and pests that destroy the crops has also been discussed in this chapter. The seaweed contains elements that also work as plant hormone regulators, bio-stimulant effects of seaweed; the use of seaweed as fertilizers and crop conditioners has also been discussed in this chapter.

5.1. INTRODUCTION

From ancient times seaweed has been used as soil additives to improve, crop productivity has been stated by researchers (Chapman and Chapman, 1980; Craigie, 2011). It is around the first half of the 1st century that seaweed was suggested for transplanting plantlets of cabbage.

A suggestion from the 4th century for the administration of seaweed at the Pomegranate roots is often alluded to it. In 1950 liquid products made from marine algae were launched, and at present being used around the world with success.

The Maxi crop has for a first came into being in the British market. Back in 1958, it was already developed for export sales and is searching for methods to acquire dry extracts, more suitable for export markets. The first extracts of dried seaweed were exported back in 1959. The purpose of seaweed fertilizer use at that time was intended to counter the conventional way of soil fertilization (Booth, 1969). They had to be sprayed when put on the market of algal extracts, which had to face the dogma of agriculture that the nutrients you needed to get into the herb via the roots, but that the leaf had no just photosynthesis ability.

The sales of products from seaweeds increased when in agriculture the foliar mineral nutrition became a regular practice in the decade of 1960. In food, and as animal feed, as a product and in the therapeutic application, marine bioactive substances derived from seaweed are in use in the industry as raw materials in present times.

The part and parcel of coastal marine habitats is seaweeds. This is primarily multicellular, macroscopic seaweeds that often occupy coastal regions of the oceans where suitable substrates occur (Khan et al., 2009).

In the agriculture the variety that is generally used is the group, the brown algae (Phaeophyta), with around 2,000 species, Ascophyllum nodosum is included in it. There are other brown algae species like Fucus spp., Laminaria spp., Sargassum spp., Turbinaria spp. that are in use in agriculture. The list of commercial seaweed products was made by Khon et al. in 2009, identifying the business that manufactures and utilizes products in agriculture.

These types of products are used as plant growth drug and also in animal food. For a long time, aquatic bioactive compounds harvested from the seaweed are used to promote plant growth and productivity. The research scientist in this field has fully capitalized on identifying new drugs and checking their efficacy. The stimulating of growth, improving the efficiency of utilization of nutrients and tolerance to different factors and biotic abiotic stress have been documented with consequences. Any progress in crop technology related to plant roots, relating to maximizing nutrient absorption, the impacts, besides reducing the harmful influence of farming on the ecosystem, increasing growth, and decreasing input into the system.

The biochemical activities of macro and microelements as well as amino acids, vitamins, and substances are responsible for the process through which these products control cellular metabolism (Crouch et al., 1992; Crouch and van Staden, 1993; Reitz and Trumble, 1996; Durand et al., 2003; Stirk et al., 2003). The complementary activity of many components at different concentrations are responsible for impact of the derivatives, hence it is still not fully clear about the mode of activity of seaweed extracts (Fornes et al., 2002; Vernieri et al., 2005).

In 2009, Khan and other researchers presented the possible mechanisms by which seaweed extracts have beneficial effects on agriculture. The enhanced photosynthesizing productivity and carbon integration, deferred senescence, antimicrobial, non-feed, and insect repulsive, decreased transpiration, increased stomatal conductivity, root exudate modulation, effective water and nutrient absorption are the various positive effects of seaweed derivatives on agriculture and the possible process is given by Khan et al. in 2009.

The application of small amounts of Ascophyllum nodosum extract on the soil or on tomato foliage induced an increase in the amount of chlorophyll in the leaves. The reduction in chlorophyll degradation that may be partially attributed to betainei from seaweed extract causes the increase in chlorophyll content (Whapham et al., 1993).

The impact of the seaweed extract products is on the growth and production of roots of plants. The derivatives of Ascophyllum nodosum assess the root growth of Arabidopsis at exceptionally low concentrations (0.1 g L^{-1}), whereas the concentrations of 1 g L^{-1} affects the height of plants and leaf numbers (Rayorath et al., 2008).

The extracts if given at an early growth stage have caused a more visible rejuvenating growth of roots. The growth in biomass has been observed by enhancing the growth of roots (Crouch and van Staden, 1993). Stimulation effects were observed both when the goods were added to the ground and when foliar was added (Finnie and Van Staden, 1985).

Seaweed concentrates cause early blooming and a large number of fruits forming (Featonby-Smith and van Staden, 1987; Arthur et al., 2003). The plant defense against pathogens and pests has been improved by use of extracts of seaweed (Featonby-Smith and Van Staden, 1983; Crouch and Van Staden, 1993; Allen et al., 2001) (Figure 5.1).

Figure 5.1. Methods of drying of the seaweed by farmers in 1940s.

Sources: Image by Wikimedia.

5.2. IMPACT OF SEAWEEDS ON AGRICULTURAL CROP PRODUCTION AS BIOFERTILIZER

The marine plants related to the Thallophyta plant kingdom are seaweeds or macroalgae (Silva, 1992; Dhargalkar et al., 2001; Kerswell, 2006; Jothinayagi and Anbazhagan, 2009; Arioli et al., 2015). This type of species is also considered an underused bioresource, though other species have been used for centuries as sources of fruit, industrial gums, and in medicinal and botanical applications (Khan et al., 2009).

There have been several researches on the stimulating impact of marine algae on plant growth (Challen and Heminway, 1965; Russo and Berlyn, 1990). The provider of antioxidants, plant growth hormones, osmoprotectants, and mineral nutrients, as well as many other organic compounds, such as novel bioactive molecules, was found to be in the seaweed (Akila and Jeyadoss, 2010; Ramarajan et al., 2013; Ismail and El-Shafay, 2015; Pacholczak et al., 2016a).

The scarcity and shortage of N, P, and K in soil was regarded to be fulfilled by the use of seaweed as biofertilizer (Sangeetha and Thevanathan, 2010; Srijaya et al., 2010; Sunarpi et al., 2010; Tuhy et al., 2015; Singh et al., 2016; Vyomendra and Kumar, 2016).

The prospect of the use of seaweeds in industrial agriculture has been extensively discussed, and different variations of these marine algae preparations are being used as liquid fertilizer and either entire or finely chopped powdered algal manures. These types of techniques like the enzyme assisted extraction, microwave assist extraction, pressurized exhaust fluid extraction, supercritical fluid extraction, and ultrasound-assisted extraction, which enable enhanced extraction without deterioration of biologically active compounds and they attracted the attention of scientist and they are now focusing on it (Michalak and Chojnacka, 2015).

The large variety of bioactive compounds is found in seaweed extracts. Their processes of extraction rely on intention. The manufacturing methods are not available in public and retained as trade secrets and it is cause behind the lack of data on algal extraction technology processes for purpose of agriculture (Craigie, 2011).

In addition, from using marine macroalgae to produce agricultural biostimulant, many extraction procedures had already been adopted. From the use of salt, alkaline, or acid methods or by chemically modifying marine algae by frying at low temperatures to micronize small particle suspension the production extracts take place (Roj et al., 2009; Sharma et al., 2014). The most cost-efficient and feasible method for enhancing the release of biomass micro and macro elements appears to be water extraction for the use of seaweed as fertilizers and biostimulant compounds (Michalak and Chojnacka, 2015).

The number of studies has thus identified the use of alkaline or water extracts from algae under natural and stress-related conditions as plant growth biostimulants like in wheat, tomatoes, Arabidopsis, and spinach, Vigna sinensis, etc. By using the process of autoclaving or heating up a previously washed seaweed in distilled water these derivatives are produced (Sivasankari et al., 2006; Kumar and Sahoo, 2011; Nabti et al., 2010; Craigie, 2011) or in alkaline solutions (Whapham et al., 1993; Rayorath et al., 2008a, b; Craigie, 2011; Fan et al., 2011).

The positive effects have been reported on growth of cereals, pulses, and flowering plants in most of the situations (Kavipriya et al., 2011). The good outcome have been recorded for promoting seed germination and root growth, improving frost tolerance, improved nutrient absorption and control of phytopathogenic fungi (Younes et al., 2009), bacteria (Kulik, 1995; Bouhlal et al., 2010; Alves et al., 2016), insects or other pests (Nassar et al., 1999; Ravikumar et al., 2011; Asha et al., 2012) and also for promoting plant growth under stressful salinity (Nabti et al., 2010; Pacholczak et al., 2016a) relying on implementation of derivatives of seaweeds.

The utilization of marine algae is encouraging and offers viable alternatives to the contamination issues created by the widespread usage and industrialization of chemical fertilizers. Though, as many writers have stated, there is an absence of full information on diet and biochemistry remediation, natural bioactive compounds, and their activities. In the key composition of seaweeds plant, growth stimulation is frequently detected by seaweed extracts, but in other cases, the stimulating processes of plant growth are not well realized.

Many researchers have displayed that the relaxing properties are attributed to a number of main components in the extracts of seaweed, such as different plant nutrients, phytohormones or betains (Karthikai Devi et al., 2009; Alam et al., 2014; Divya et al., 2015a; Michalak and Chojnacka, 2015; Shahbazi et al., 2015; Al-Hameedawi, 2016; Mirparsa et al., 2016). In the other side, because of its nutrient content, organic matter found in seaweed biofertilizers is considered to promote plant growth (Davari et al., 2012).

It has also been shown that the introduction of sufficient amounts of various seaweeds as organic fertilizers increased soil quality and fertility specifications of field crops (Badar et al., 2015). Seaweeds increase soil nutrient levels such as N, P, and K, and other minerals required to grow plants (Imbamba, 1972; Tay et al., 1987; Sethi, 2012; Mirparsa et al., 2016).

The mineral composition tests in different seaweeds belonging to separate taxonomic classes such as Chlorophyceae (Chladophora glomerata, Ulva reticulata, Halimeda macroloba) and Rhodophyceae (Gelidiella acerosa, Hypenea musciformis) showed the quality of various minerals such as F, Ca, Ag, Pa, K, Fa, Mn, Zn, Cu, Ni, Co, Cr, Cd (Anantharaman et al., 2010; El-Said and El-Sikaily, 2013; Tuhy et al., 2015).

In addition, marine algae are high in protein, sugar, starch, cellulose, hemicellulose, lignin, vitamins, bromine, and iodine (Staden et al., 1994; MacArtain et al., 2007; Crouch and Van Staden, 1993; Mohammadi et al., 2013; Shri Devi and Paul, 2014; Heltan et al., 2015; Michalak et al., 2015; Mirparsa et al., 2016). They also cover a wide range of organic compounds including at minimum seventeen amino acids (Qasmi, 1991; Oohusa, 1993; Yoo, 2003; Shevchenko et al., 2007).

In marine seaweed, a higher concentration of minerals was observed in contrast with land vegetables (Manivannan et al., 2008; Kumar et al., 2009). Aslem and colleagues analyzed the mineral composition of a red algal extract (Lithothatmnion calcareum) and demonstrated that the mineral abundance contributed to the utilization of the algal extract as animal and human food (Aslam et al., 2010).

In numerous marine algae a high abundance and quantity of polysaccharides is exceptional. Such carbohydrates enter cell walls and others are within the cell storage compounds (Murata and Nakazoe, 2001; Eldeek and Mervat, 2009; Cian et al., 2015; Heltan et al., 2015). The various polysaccharides of seaweeds typically have a chemical composition similar to the algae's related taxonomic classification (Chojnacka et al., 2012).

The high content of polysaccharides means a high degree of soluble, insoluble fibers. Chemical structure varies with the seasons and under various environmental factors, such as temperature, salinity, light, and supply of nutrients (Reeta, 1993; Marinho-Soriano et al., 2006; Karthikai Devi et al., 2009; Anantharaman et al., 2010; Hanan and Shimaa, 2013).

The quantity of soluble, insoluble, and dietary fiber in both Ulva pertusa and Ulva intestinalis species varied in weighing in dry condition from 25.0–39.6%, 21.8–33.5%, and 51.3–62.2%, respectively (Benjama and Masniyom, 2011). The significant nutrient composition difference occurred between cold and rainy winter and warm summer season (Benjama and Masniyom, 2011).

The large quantities of lipids, phenols, chlorophyll a and b and complete chlorophyll were observed in various species of the genus Ulva, especially U. Rigida (Satpati and Pal, 2011) and U. Lactuca (Abd El-Baky et al., 2008). The abundance of pigments (chlorophyll, caroteinoids, and phycobiliproteins) has also been found of addition to carbohydrates (Chojnacka et al., 2012).

5.3. SEAWEEDS AS BIOFERTILIZERS

The seaweeds were used as soil additives to increase production of crops in coastal regions and for the regeneration of alkaline soils where nutrient deficiency is widespread from ancient times (Zodape, 2001; Craigie, 2011). The manure made from seaweed has been used in large areas of lands in the second half of the 1st century (Metting et al., 1990).

The seaweeds were combined with or composted with sand and soil or with grass, feeding, or other agricultural waste (Craigie, 2011). The observed advantages of seaweed as a renewable bioactive and organic matter supply and fertilizer nutrients contributed to a broader use in the farming sector (Chapman, 1980; Nelson and Van Staden, 1984).

In order to fertilize soil few brown seaweeds have been used as it contain alginate in large quantity that is broken down by soil microbes as an added organic matter (Thivy, 1964). The micronutrients from different seaweeds have recently been applied to a growing volume of plant biomass owing to the high levels of zinc (Tuhy et al., 2015).

As a nutrient-rich natural organic fertilizer, the new generation of seaweed liquid fertilizers (SLF) has been developed, which encourages faster growth of seed, increases crop yields, and enhances pathogenic resistance of several crops (Sathya et al., 2010). In agriculture and horticulture there, is a large-scale application of LF that are based on the extracts of algae that was first thought and developed by Milton in 1952 (Hurtado et al., 2009; Srijaya et al., 2010; Shahbazi et al., 2015; Ciepiela et al., 2016).

The uses of seaweed as biofertilizers are as follows:

The application of marine macroalgae is also for soil stabilizers and not only for the use as biofertilizers (Abdel Raouf et al., 2012; Bhardwaj et al., 2014; Arioli et al., 2015). There is a particularly good impact of fresh kelp (Macrocystis integrifolia) on crop growth and nutrition that are planted in fine textured soil (Temple and Bomke, 1988).

The fertilizers extracted from some genus of seaweeds (Fucus, Laminaria, Ascophyllum, Sargassum) are bio-degradable, non-toxic, non-polluting, and non-harmful to humans, animals, and birds (Dhargalkar and Pereira, 2005). The Nedzarek and Rakusa-Suszczewski (2004) demonstrated that high amounts of organic matter and various nutrients, that are particularly rich in NH_4, NO_3 and NO_2 and phosphates, have been released from macroalgae mixture.

In addition, growth stimulation of okra was found after foliar application (Abbasi et al., 2010), stimulation of Vigna mungo by extracts of Sargassum myriocystum (Kalaivanan and Venkatesalu, 2012), induction of amylase activity in barley (Rayorath et al., 2008a), effect of Ulva fasciata on seed germination, growth parameters, pigment, and carbohydrate content (Shahbazi et al., 2015), enhancing rice and maize growth from seaweed sap (Singh et al., 2015a, b, 2016), and enhancing root and shoot length with increased leaf numbers.

Most recently, Vigna spp. overall growth stimulation by utilization of macroalgae is discovered as a biofertilizer (Reddy et al., 2016; Vyomendra and Kumar, 2016; Binumol and Gopalakrishnan, 2016). Ascophyllum nodosome commercial extract therapy impaired the regulation of the phytohormone biosynthesis and aggregation in Arabidopsis (Wally et al., 2012).

Also, the use of brown macroalgae derivatives Ascophyllum nodosum (Khan et al., 2011) has demonstrated induction of cyto-kinina-like behavior in Arabidopsis thaliana. Many origins of biofertilizers have been explored in several ways (Reitz and Trumble, 1996; Kavipriya et al., 2011; Lola-luz et al., 2013; Pramanick et al., 2013; Zamani et al., 2013; Jayasinghe et al., 2016) (Figure 5.2).

Figure 5.2. Farmer collecting the seaweed for use on farm.

Sources: Image by Peakpx.

5.4. EFFECTS ON DISEASES AND INSECT PESTS

The powdery mildew spreading from the former leaves and that of the gray molt (Botrytis cinerea) on stem fruit was prevented by the alkaline hydrolyzed extract, Maxicrop, sprayed weekly in turbot and strawberry. The both crops produced increased yields (Stephenson, 1966). The tomato planting failure owing to damping was substantially reduced when watered weekly with hydrolyzed extract 1:120 dilution.

The latest research study has measured the impact on two strawberry crops (Elkat and Salut, 2004) in Poland of liquefied seaweed products Goëmar BM 86 and Kelpak SL (Masny et al., 2004). For each, two-year field trials have shown that the marketable Elkat fruit yield of Goëmar BM86 increased respectively by 17 and 43% for the 1- and 2-years old plants; no effect was observed on either cultivar's fruit weight.

The production of strawberry with the salut cultivar is not affected by any seaweed product, while the fruit weight was decreased by Kelpak SL, and not in the cultivar Elkat. In each of these cultivars, these two-seaweed spray decreased fruit strength. Even if B. Cinerea affected the fruit up to 24% and the occurrence of gray mold was not decreased by any solution made from seaweed. Obviously, varietal variations are of importance in a crop's response, and the quality of the seaweed extract is itself a variable (Mansy et al., 2004).

The product Goëmar was made from A. nodosum as Kelpak was from Ecklonia sp. and made from A, with the exception of Maxicrop. Both have been manufactured using technology of cell explosion. Through the assessment of relative bioefficacy of marine extracts, processing methods should therefore be recognized.

The treatment with a plant systemic inducer or elicitor, such as salicylic acid, jasmonic acid, arachidonic acid, seaweed extract and humic acid as well as reactive oxygen donor, enhances pathogenesis proteins (PR1, phenylalanine ammonium lyase PAL, or rich hydroxyprotein) to ensure disease protection (Moon and Anderson, 2003, 2006).

A fluid fertilizer A. nodosum (GYFA 17, Goëmar) contains laminaran that was shown to upright for use in tobacco as defensive responses for PAL, O-methyl trans-caphaelic acid, lipoxygenase, and salicylic acids (Klarzynski et al., 2000; Patier et al., 1993; Potin et al., 1999), and anti-fungal compounds (Kobayashi et al., 1993). The powerful elicitor responses have been reported in Ț-carrageenan-treated tobacco leaves (Mercier et al., 2001).

The seedlings threatened with Xanthomonas campestris developed strong bacterial pathogen resistance, when the seeds were soaked in the aqueous formulation Sargassum wightii extract 1:500 Dravyas solution for 12 hours prior to germination (Raghavendra et al., 2007). The production of cotton seedlings sprayed with marine extract showed improved resistance to bacterial attack, where the greatest reduction in the amount of blight that is up to 74% after 80 days occurred when the seed treatment was combined with a spraying protocol.

The evidence from peroxidase measurements and increased total phenolic levels in treated plants showed systemic resistance. The extract of algae was superior to streptocycline care by encouraging more plant vigor, increased stem rotation, and increased boll yield without the risk of releasing an antibiotic into the environment. In chloroform extracts of red and brown seaweed, alcohols, and polyunsaturated esters have been used as antibacterial agents (Vallinayagam et al., 2009).

The low levels of the glycine betaine known to be present in seaweed extracts give Uromyces phaseoli resistance in beans (Blunden et al., 2009).

The creeping bent grass showed an increased dismutase of superoxides and reduced dollar spot disease in a treatment (Sclerotinia homoeocarpa). The derivative of A. nodosum and humic acids (Zhang et al., 2003).

The increases in the levels of phenolic ingredients, flavonoids, and antioxidants in spinach leaves watered with soluble A have been identified and that is extract of A. nodosum (Acadian Sea plants) (Fan et al., 2010). The Stephenson in 1966 stated that the black bean aphid infestations on small bean leaves sprayed with Maxicrop were greatly reduced in comparison with water sprayed controls. She further remarked that fewer winged adults settled on extract sprayed sugar beet leaves than on the controls that suggested that the aphides had an aversion rather than an insecticide effect.

In the same manner, on trees sprinkled with hydrolyzed algae extract, the population of red spider mites (Tetranychus telarius) in orchards has been kept to low levels in comparison to the non-sprinkled plants. When untreated leaves were picked, the mites seemed to deliberately resist settling on extract-treated apple leaves.

The reduction of spider mites and eggs sprayed with Maxicrop on glasshouse chrysanthemums was greater than that of controls receiving the standard acaricide protocol. In the strawberries grown in high polyethylene tubes, two-spotted red spiders (Tetranychus urticae) populations have decreasing dramatically after two weeks of spray with Maxicrop Triple, which basically confirms the prior observations (Hankins and Hockey, 1990).

While chelated metals on the leaves can reduce the fertility of two-spotted spider mites (Terriere and Rajadhyaksha, 1964), the mechanism of action remains unknown. Increased levels of anthocyanins and phenolic ingredients in the leaves can change the palatability or acceptability of insect predators' leaves. Seaweed extract Kelpak 66, a predator root of nematode (Meloidogyne incognita) in tomatoes is stated to be reduced by 1:500 (Featonby-Smith and van Staden, 1983) either as a foliage or as a soil drench.

At transplantation, a single soil dunk was the greatest change in the roots, leaves, and fruits of heavily infected plants. Of course, when the algae extract was added the nematode population in the soil grew beyond level, but the number of nematodes recovered from within the roots was much lower than that of the infected level.

The reduction of root-knot nematode fecundity M. javanica occurred with in vitro inoculation of Arabidopsis seedlings with juvenile nematodes when the seedlings were handled with Maxicrop or an equal quantity of a betanium mixture (Wu et al., 1998). Kahn et al. (2009) addressed the alleviation of biotic stress by algae products.

5.5. PLANT HORMONES REGULATORS

The modern concept of plant hormones requires that they be naturally occurring organic substances which influence physiological processes at low concentrations, far below those where nutrients or vitamins affect those processes (Davies, 2004). There is no requirement that the hormone be transported within the plant. A plant growth regulator is a term commonly used in the agrochemical industry to distinguish synthetic plant growth regulators from the endogenous ones. Following the first report of auxin in marine algae (van Overbeek, 1940), attention focused on plant growth regulators that could be detected in algae.

The definite identification of IAA, however, remained questionable until the 1970s although Avena coleoptile bioassays revealed several indole type compounds released by alkaline hydrolysis of several seaweeds including Fucus vesiculosus and A. nodosum (Buggeln and Craigie, 1971). It is now recognized that many of the common higher plant hormones (abscisic acid (ABA), auxins, cytokinins (CKs)) occur in algae including brown seaweeds (Jameson, 1993; Tarakhovskaya et al., 2007; Zhang et al., 1993).

The other growth regulators also are reported from brown algae. The presence of IAA, phenylacetic acid and its para-hydroxy derivative were reported for Undaria pinnatifida using chromatographic, IR, and MS techniques (Abe et al., 1972, 1974), but the quantity of IAA found in the harvested seaweed was small ($\sim$10 μg kg−1 wet weight). The commercial extracts Maxicrop (A. nodosum) and Kelpak 66 (E. maxima) were examined by GC-MS and found to contain IAA (Crouch et al., 1992; Sanderson et al., 1987).

The cell burst concentrates of both E. maxima and M. pyrifera contained indole-amino acid and other indole conjugates in addition to IAA (Stirk and van Staden, 2004). The auxin content, however, declined following a fortnight of storage at 54°C. The ABA is widely distributed in algae being detected in 64 species from nine divisions by ELISA (Hirsch et al., 1989).

In this of particular interest is its occurrence in ten genera of brown algae including A. nodosum, a commercial extract of it, and E. maxima and Laminaria spp. (Boyer and Dougherty, 1988; Hirsch et al., 1989). The function of ABA in the algae remains to be discovered although it induces genes for protein synthesis required for drought tolerance of seeds in terrestrial plants (Rensing et al., 2008; Verslues et al., 2006).

The ABA induces stomatal closure in drought, protein storage in seeds, and gene transcription for proteinase inhibitors, inhibits shoot growth, and induces some seed dormancy effects (Davies, 2004). The evidence for ethylene production in brown seaweeds is tenuous although 1-aminocyclopropane-1-carboxylic acid, a potential ethylene precursor, was reported in Kelpak 66, a commercial extract of E. maxima (Nelson and van Staden, 1985).

The results from paper chromatography and a dwarf pea bioassay indicated that gibberellin-like substances (possibly GA1, GA3, and GA6) occur in F. vesiculosus (Radley, 1961). The gibberellin-like bioactivity was observed in a lettuce hypocotyl elongation assay of freshly prepared commercial extracts of Ascophyllum; though, the activity was unstable and disappeared within 4 months of storage (Williams et al., 1981).

Direct chemical evidence for gibberellins does not appear to exist for brown algae, although traces of GA4 were identified by MS in a soluble extract produced by Acadian Seaplants Limited from A. nodosum (R.P. Pharis, pers. comm.). Cytokinin-like bioactivity was reported in the early 1970s in commercial seaweed extracts (Williams et al., 1981).

The experimental trials with these extracts resulted in increased potato yields, elevated protein content of grass, and a pronounced retardation in the loss of green color in ripening lime fruits, effects attributed to cytokinin activities of the extracts (Blunden, 1977). CKs and zeatin riboside are reported to improve heat tolerance in creeping bent grass (Zhang et al., 2010).

The evidence that CKs are widely distributed in seaweeds was presented by Stirk and others in 2003 and they used immunoaffinity chromatography, HPLC-MS, and single ion measurements to investigate five green, seven brown, and 19 red species of seaweeds. All 31 species contained isoprenoid CKs, principally cisand trans-zeatins, their ribosides and O-glucosides, and isopentenyl adenine and -adenosine.

The quantitative measurements referenced to deuterated cytokinin standards revealed that cis-zeatins generally predominated over the transzeatins. In addition, benzyladenine was present in all species but at concentrations usually lower than those of its ortho- and meta-hydroxylated derivatives (topolins). It is interesting to note that the topolins can act as substrates for zeatin-O-glucosyl transferases (Mok et al., 2005), and that trans-ZOG and meta-topolin are dominant CKs in cell burst seaweed concentrates (Stirk et al., 2004).

The qualitative and quantitative changes relating to seasonal and lunar cycles were reported for endogenous CKs in E. maxima, and Sargassum heterophyllum (Featonby and van Staden, 1984; reviewed by Jameson, 1993). The cytokinin concentrations in S. heterophyllum also were greatest at the onset of vegetative growth and during gamete production (Mooney and van Staden, 1984). The CKs in M. pyrifera were most concentrated (23.4 ng g−1 fresh weight (FW)) in the intercalary meristem of young blades (de Nys, 1990).

The cytokinin metabolism is evolutionarily old as the signaling molecules (N6-substituted adenine derivatives) occur in bacteria (Barciszewski et al., 2007); however, the structure of the active sites of the histidine phosphotransfer (HPt) proteins differs from that of higher plants (Sugawara et al., 2005).

At least one gene coding for HPts and a regulatory protein involved in a cytokinin feedback signaling system have been reported in the green algae O. tauri, Chlamydomonas reinhardtii, and Volvox carteri (Pils and Heyl, 2009). By comparison, there were two HPts in rice, ten in poplar, and five in Arabidopsis. Nuclear HPt activates response regulators (RRs) which are involved in negative feedback regulation of cytokinin signaling pathways.

There are two to four RRs in the algae, 7, or 14 in mosses, 23 in rice, 32 in poplar, and 22 in Arabidopsis. However, proteins of the CHASE domain responsible for binding a cytokinin ligand to a receptor were not detected in the algae, although two to five were identified in the other species (Pils and Heyl, 2009). Betaines, while not traditionally included among the classical plant hormones, are widely distributed in plants including seaweeds and are found in seaweed extracts (Blunden et al., 1982, 1984, 1986; MacKinnon et al., 2010).

The glycine betaine, γ-amino butyric acid betaine, δ-aminovaleric acid betaine, and laminine have been reported in A. nodosum and F. serratus (Blunden et al., 1984). These N-methylated compounds are good potential sources of formaldehyde in biological systems and have been implicated in a double-immune response of beans to bean rust, Uromyces phaseoli (Tyihák, 2006). The concentrations of glycine betaine as low as 10−12 M seemingly.

5.5.1. Phytohormones Content of Seaweeds

The seaweeds and their extracts have been used for centuries in agriculture to improve plant growth and stress tolerance. It was concluded that their phytohormonal compounds may modulate innate pathways for phytohormone biosynthesis in plants (Wally et al., 2012; Divya et al., 2015a, b).

The seaweed extracts are known to promote and enhance vegetables, fruits, and various other crops because of their richness in growth regulators such as auxins (IAA and IBA), gibberellins, and cytokinin, in addition to osmoprotectant betains and micronutrients (Ghoul et al., 1995; Renuka Bai et al., 2007; Divya et al., 2015a, b; Mathur et al., 2015; Pacholczak et al., 2016b).

The LF based on different seaweed extracts are useful for achieving higher agricultural production, because the extract contains growth-promoting hormones, like CKs or gibberellins—in addition, to trace elements, vitamins, amino acids and antibiotics (Craigie et al., 2007; Sathya et al., 2010; Wally et al., 2012; Divya et al., 2015a; Mirparsa et al., 2016). Also, Zodape et al. (2010) concluded in their studies that the effect of seaweed extract is due to plant growth regulators such as CKs. Ascophyllum nodosum extracts have been used as biostimulant to promote growth and productivity in a number of agricultural production systems (Rayorath et al., 2008b).

It is well established that gibberellic acid (GA3) is responsible for the stimulation of seed germination of different plants (Brink and Cooper, 1947). This substance induces hydrolytic enzymes in the aleurone layer surrounding the endosperm (Cooper and Brink, 1945; Scott et al., 1998).

Jennings (1968) confirmed that green and brown algae are extraordinarily rich in GA3, which is involved in seed germination. In fact, GA3 acts as a signal in the germinating process of seeds by activating amylase genes in aleurone cells (Sun and Gubler, 2004). Rayorath et al. (2008a) reported that the organic components of Ascophyllum nodosum induce GA3-independent amylase activity in barley.

The analyzed Ascophyllum nodosum extracts revealed high concentrations of CKs, particularly trans-zeatin-type CK, and ABA (Wally et al., 2012). Since many phytohormones stimulate root development, the reported increased plant growth and vigor after application of seaweeds may be through increased efficiency of nutrients and water uptake (Russo and Berlyn, 1990).

Thus, seaweed cultivation and its utilization are an economically successful approach in agricultural production (Sridhar and Rengasmay,

2011; Gireesh et al., 2011; Michalak et al., 2016a). The seaweed extracts stimulate various aspects of plant development (Pramanick et al., 2013; Ismail and El-Shafay, 2015).

The drought generates oxidative stress and increased cell membrane leakage in stressed wheat plants resulting in an increased need for antioxidants (enzymatic and non-enzymatic) activities. The physiological effects of the application of seaweeds during drought on Triticum aestivum were evaluated in the presence and absence of different seaweed extracts (Sargassum latifolium, Ulva lactuca and their mixture) (Kasim et al., 2015).

A pretreatment with seaweed extract led to a stimulation of antioxidant activities and thus to the alleviation of the damaging effects of drought on the vegetative stage of Triticum aestivum. Sargassum or Ulva extracts antagonize the oxidative damaging effects of drought not only directly through activating the antioxidative system (catalase (CAT), peroxidase, and ascorbate), but also through providing phytohormones and micro-nutrients essential for wheat growth (Kasim et al., 2015).

When seeds of Phaseolus aureus were treated with extracts from the red marine algae Asparagopsis taxiformis, a promotion of root and internode length of shoot, as well as on petiole and leaf surface area (Renuka Bai et al., 2007). These researchers proposed that plant growth stimulation by extracts of these red algae is due to high quantities of CKs.

Many studies concerning the composition of liquid seaweed fertilizers (LSF) demonstrated that they are rich in plant growth regulators such as the auxin indole acetic acid (IAA), kinetin, zeatin, gibberellin, CKs, ABA, ethylene, in addition to betaines and polyamines (Kingman and Moore, 1982; Nelson and Van Staden, 1985; Crouch and Van Staden, 1993; Zahid, 1999; Tarakhovskaya et al., 2007; Zhang and Ervin, 2008; Zodape et al., 2008).

According to Savasangari et al. (2011), the growth enhancement of the cluster plant Cyamopsis tetragonolaba is due to the presence of cytokinin and magnesium, which are considered as an essential growth-promoting hormone and chief constituent in chlorophyll biosynthesis, respectively. The same explanation about the richness in phytohormones (auxin and CKs) in seaweeds LF (Sridhar and Rengasamy, 2010).

In the same manner, Stirk et al. (2014) showed that Ecklonia maxima extracts present a high potential for application in agriculture due to its high content of several plant hormones such as ABA, gibberelins, brassinosteroids, and castasterone.

It was also observed that the enhancement of germination, growth, and productivity of brinjal by using Sargassum wightii liquid fertilizer is related to the presence of high levels of several phytohormones in this extract (Divya et al., 2015b). The liquid seaweed extracts also have been successfully applied to restore plant growth at hardness conditions such as high pH and temperature conditions (Briceño-Domínguez et al., 2014; Bradacova et al., 2016) as well as at very low pH and water content (Arthur et al., 2013).

It was commonly reported in these studies that the main effect of seaweeds extracts against environmental stresses is due to CKs as well as gibberellic and ABA, which help to support plant growth under nutrient-stressed conditions and to recover plants after damage (Reitz and Trumble, 1996; El Shoubaky and Salem, 2016).

5.6. BIOSTIMULANT EFFECTS OF SEAWEED ON PLANTS

The liquid extracts of brown and green algae are being sold as biostimulants or biofertilizers in different brands (Kavipriya et al., 2011; Sharma et al., 2012; Divya et al., 2015a; Shahbazi et al., 2015; Gopalakrishnan and Binumol, 2016; Mirparsa et al., 2016). Algae, particularly seaweeds, are used as biofertilizers, resulting in less runoff of nitrogen and phosphorus fertilizers after application of livestock manure (Abdel-Raouf et al., 2012), which could be due to stimulated nutrient uptake in crop plants.

Many results were obtained with seaweed extracts such as improvements of seed germination and root development, frost resistance, nutrients uptake, increased resistance to pathogenic fungi, reduced incidence of insect attack and salt stress (Abetz, 1980; Nabti et al., 2010; Battacharyya et al., 2015). Adequate amounts of potassium, nitrogen, micronutrients, humic acid, polysaccharides laminarin, alginates, and carrageenans, in addition to growth-promoting phytohormones, present in seaweeds, makes them excellent biofertilizers (Sathya et al., 2010; du Jardin, 2015).

The application of seaweed saps in green gram (Pramanick et al., 2013) resulted in growth and yield increases. An increase in chlorophyll content was achieved in different crops (tomato, dwarf French bean, wheat, barley, and maize) treated with Ascophyllum nodosum alkaline extracts. The authors explained this increase in chlorophyll content by the presence of betaines in the seaweed extracts (Blunden et al., 1997).

In addition, when Ecklonia maxima extracts were applied as a foliar spray or a root drench for promotion of growth and development of marigold Tagetes patula, significative enhancement of both vegetative and reproductive growth could be demonstrated (Staden et al., 1994). A seaweed-based liquid fertilizer of Rosenvingea intricata was used successfully to increase growth and pigment concentration in Cyamopsis tetrogonolaba and Abelmoschus esculentus (Thirumaran et al., 2009a, b).

A growth stimulation of egg plants (Solanum melongena) by Ascophyllum nodosum extracts (Bozorgi, 2012) and enhancement of seed germination of fenugreek by seaweed extracts were reported (Sabale and Pise, 2010). Sridhar and Rengasamy in 2010 successfully applied Sargassum wightii (brown seaweed) mixed with Ulva lactuca (green marine alga) to increase peanut growth (Arachis hypogaea). Additionally, a combination of Azotobacter chrocoocum and Bacillus megaterium var. phosphaticum mixed with marine algal extracts stimulated growth of bitter orange seedlings (Ismail et al., 2011).

Furthermore, Azospirillum brasilense strain NH combined with Ulva lactuca extracts was able to enhance the development of durum wheat under saline and non-saline conditions (Nabti et al., 2010). Moreover, a commercial algal extract of Ascophyllum nodosum was proven as potential bioregulator which improved the size and quality of fruits and reduced variations in fruit yield during different seasons in apple trees (Spinelli et al., 2009).

The latest studies also showed that some algal extracts exert potent biostimulation of vegetative growth and yield of fig trees (Ficus carica) (AlHameedawi, 2016). The application of a variety of LSF were repeatedly reported to be able to decrease application doses of nitrogen, phosphorus, and potassium as well as to stimulate plant growth and yields of many crop plants. There are at least 59 species of seaweeds that were shown to stimulate germination, growth, and yields of some horticultural and legume plants (Sunarpi et al., 2010).

He obtained satisfactory results when using a mixture of ten seaweeds to improve growth and production of rice. Seaweeds are much better than chemical fertilizer to improve soil fertility, because they contain a high quantity of organic matter which additionally improves available water and mineral uptake in the upper soil level (Ramarajan et al., 2012).

SLF of Sargassum wightii, Ulva lactuca, and Enteromorpha intestinalis enhanced seed germination, growth, and biochemical features of legume plant Glycine max (Mathur et al., 2015). These extracts were shown to induce seed germination and growth parameters more strongly than chemical fertilizers (Ramarajan et al., 2012; Godlewska et al., 2016). Finally, Partani (2013) showed an extraordinary effect of seaweed extract on corn growth and performance.

5.7. SEAWEED AS FERTILIZERS AND SOIL CONDITIONERS

The majority of seaweed found in nature in large quantities is brown seaweed and is used to prepare the seaweed meal by drying and crushing it. The popular varieties are Ascophyllum, Ecklonia, and Fucus. It is sold as soil additives and act as fertilizers as well as soil conditioners.

They are appropriate for nitrogen and potassium subject matter, however in phosphorus they are so much different than conventional manures and are typically in chemical fertilizers with N: P: K ratios. In brown sea algae, the large quantities of insoluble carbohydrates work as soil conditioners and improve aeration and soils, particularly in clay soils.

The soils retain good humidity due to it. Their usefulness as fertilizers is often due to the trace elements. In contrast to standard plant specifications, the real impact they provide is quite limited. Maerl is the fertilizer produced by red seaweed Phymatolithon alcareum and Lithothamnion corallioids, which grow with a calcium carbonate crust on top.

It is located primarily at the coasts of the French coast, next to the sources of rivers and calm bays, in which the water temperature would be 13°C or higher. They are collected by dragging or digging and used as a replacement for agricultural lime to neutralize acid soils.

Maerl is costly rather than lime, although its trace elements are expected to be more cost effective, although trace elements can be incorporated more easily (Figure 5.3).

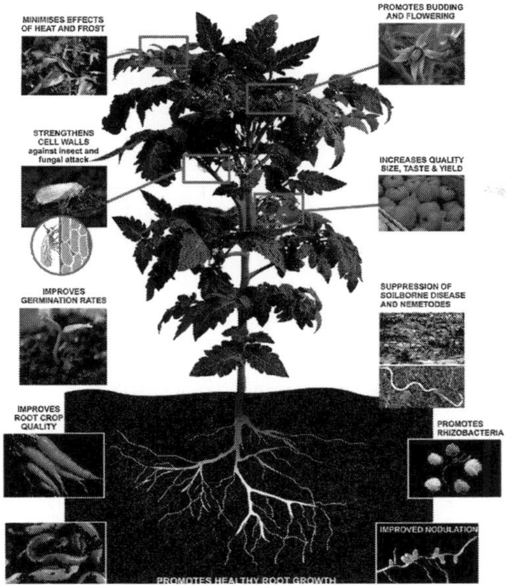

Figure 5.3. Benefits of seaweed fertilizer.

Sources: Image by Wikimedia.

5.8. USE OF SARGASSUM AS CROP ENHANCER

Crops seaweed extracts are important in crop production and are used in several ways, such as foliar applications and soil drenches during transplantation (Mohanty et al., 2013) as well as seed treatments (Kalaivanan and Venkatesalu, 2012). The extracts allow for variations in concentration of the product; are also easy to dilute and tend to have rapid effects on targeted crop species. Although liquid extracts are the most commonly used seaweed product, application, concentration, and extraction methods have not been standardized (Mohanty et al., 2013).

The evidence suggests that the effectiveness of foliar seaweed extract sprays on plants is dependent on the concentration of the seaweed. For instance, foliar sprays with 20%Sargassum spp. extracts compared to other concentrations (0%, 10%, 50%, 100%) resulted in increased growth, yield, and improvements in the quality of tomato plants (Sutharsan et al., 2014).

In critical food crops such as legumes, seaweed extract improved the vigor and productivity of legumes in the midst of harsh soil conditions (Salma et al., 2014). Also, with its rich content of zinc, copper, and manganese, seaweed extract application has impacted positively on fruit formation in tomato (Sutharsan et al., 2014). Globally, seaweed is commonly used as manure in coastal areas (Kumar et al., 2012).

Additionally, Sargassum spp. based fertilizers used within different crop production systems have resulted in increased crop productivity (N'Yeurt and Iese, 2014; Crouch and Van Staden, 1993). Seaweed, according to Salma et al. (2014), contains major and minor plant nutrients; organic compounds such as auxins, gibberellins, and precursors such as ethylene and betaine that once applied appropriately, has the potential to stimulate plant growth. For instance, Crouch, and Van Staden (1993) in their studies observed faster fruit ripening rates, a 17% increase in the weight of fresh fruit and approximately 10% increase in the amount of fruits harvested.

According to Chapman and Chapman (1980) as cited by Simoons (1991), their high potash content makes them quite suitable for fertilizing root crops and other crops. For instance, in parts of China, Sargassum spp. is seen as a low-cost fertilizer in sweet potato production (Simoons, 1991). In the application of seaweed in the raw or composted form, soil salinity is a major issue.

The composts produced from plants originating from sea sources such as Sargassum spp. seaweeds are known to have a high salt content and therefore limits potential use as composts in agricultural crop production (Vallini et al., 1993). The composting requires that excessive salts be removed either prior to/or during the composting process. Due to the high salt and sand content of raw material that is seaweed, (Eyras et al., 1998) had to alter their composting process and added water periodically to reduce salts and sand.

The process of removing the excessive salt and sand can be expensive (Vallini et al., 1993) if consideration is made that freshwater is needed to remove the salts. The use of water to wash out excessive salts may also be counter-productive due to the washing out of beneficial nutrients that can be found in the seaweed compost.

5.9. SOIL CONDITIONER IN ORGANIC FARMING

In recent years, seaweed extracts have acquired a substantial global appeal as LF (Khan et al., 2009; Kumar and Sahoo, 2011; Shah et al., 2013), while they have been used widely as an agricultural biofertilizer since its initial use as soil conditioner for horticultural crops (Aitken and Senn, 1964). Seaweeds currently have foliar spray, granules, powder, and manure in their application as a fertilizer (Kumari et al., 2011, 2013; Gharakhani et al., 2016).

The latest developments in organic agriculture and growing demand for organic foods have provided a wide range of opportunities for marine researches as a soil conditioner, as a biofertilizer, as well as biostimulant for their future applications (Khan et al., 2009).

Seaweeds include a variety of organic compounds considered to be helpful in the conservation of hydrogen and nutrients in the upper eyes in particular, vitamins, proteins, and fatty acids. Many researchers have demonstrated the beneficial impact of seaweed dependent fertilizers on land and crops by growing the potential for water preservation, stimulating production of microorganisms, and improving soil texture.

Seaweed fertilizers allow the soil to establish a root growth atmosphere by growing microbial diversity and promoting biological operation such as breathing and activation of nitrogen and mineralizing mineral nutrients (Haslam and Hopkins, 1996; Selvaraj et al., 2004; Battacharyya et al., 2015) (Figure 5.4).

Figure 5.4. Soil conditioner made by crushing the seaweeds.

Sources: Image by Flickr.

5.10. SEAWEED LIQUID FERTILIZER (SLF)

The trace elements, macro-nutrients, organic compounds such as amino acids and regulators for plants' growth, like cytokinin in, auxin, and gibberellin are found in SLF. According to Verkleij (1992), the ability of the soil to hold water is enhanced by the application of SLF. The large amounts of chemical fertilizers to make up for soil nutrient deficiency are done by farmers all over the world.

The abundant use of chemical fertilizers is noted affecting soil and plants over the period of time. The beneficial effects on physical and chemical soil properties of organic material as supply for many of the nutrients needed, that will eventually improve productivity is presented by Singh and Yadav (1992).

The importance of seaweeds as fertilizer is attributed not just to nitrogen, phosphorus, and potash, though due to the existence of trace elements and plant growth regulators like metabolites (Booth, 1969). The SLF, LSF, LF (liquid fertilizers) are different products made from seaweed.

The algal seems to be either entirely or eventually cut powdered and has all been noted to have beneficial effects on cereals, pulses, and flowering plants. The benefit of marine manures is that they are independent from weeds and pathogenic fungi. The usage of SLF offered an improved crop yield, stress tolerance, uptake of nutrients, stronger germination of the seeds, and decreased occurrence of fungal attacks. The significant amounts of plant growth regulators have been known to found in seaweed.

Seaweeds are known to contain substantial quantities of plant growth regulators like substances (Mooney and Van Staden, 1985), Cytokinin (Smith and Van Staden, 1984), IAA (Abe et al., 1972), gibberelins, and gibberellin (Bentley, 1960; Radley, 1961; Sekar et al., 1995). The excessive use of chemical fertilizers and pesticides is therefore a matter of great importance, particularly in the third world, for marine algae, especially marine algae, are of vital importance in agriculture.

The derivative from seaweed has been found in a number of plant species to improve stress tolerance. The result was a higher concentration of bioactive molecules in treated plants (Rayorath et al., 2009; Zhang and Schmidt, 2000) like antioxidants.

In deciding environmental limits on biological activity, the exhaustive ground level test is especially important with the drug, and for controlling inoculum survival and dispersal (Davison, 1988). The viable solution for the long term is therefore the use of modern farming in accordance with traditional agriculture practices.

5.11. EFFECT OF SEAWEED EXTRACT ON CROP PRODUCTIVITY

It has been discovered that seaweed fertilizer is better compared to chemical fertilizer. The availability of high organic matter in the upper soil leads to the preservation of moisture and minerals required for roots (Sivasankari et al., 2006). The Arthur (2003) and Zodape et al. (2008) recorded substantial improvements in crop yield because of the foliar application of marine weed derivatives.

The useful strategy for enhancing production of rainfed soybean may be the foliar application of seaweed extracts (Rathore et al., 2009). The use of marine algae also significantly improves oil content, oil yield, K, Na, and sunflower seed crude protein. The application of 0.6% Gracilaria dendroides concentrations and Ulva lactuca concentrations is recorded to have been higher increase in sunflower oil by 34.05 and 30.55% (Hannan and Salem, 2011).

The highest production of straw with 15% seaweed extract was also accomplished in the context of green gram (Pramanick et al., 2013). In comparison, applications of 38.97 and 33.58% higher grain yield than controls in the green gram, along with the prescribed amount of the fertilizer, were recorded with 15% Kappaphycus spp. and the prescribed dose of the fertilizer (RDF) accompanied with Gracilaria spp. and the prescribed dose of fertilizer (REDF).

In Kappaphycus and Gracilaria extracts the existence of micro-elements and plant growth regulators, particularly CKs (Zodape et al., 2009), is cause for the increase in yield and enhanced green gram nutrition, which is foliarized by the above two saps. The usage of seaweed extract enhanced the early growth and yield attribute in leguminous plants and increased the yields of the algae 12–25% by 47.52% and 42.52% by 10% of Kappaphycus Alvarezii and Gracilaria edulis extracts compared to the application of Gram 42.52 respectively.

Owing to the foliar spray of sea weed, the average production of beans was increased by 25% (Temple and Bomke, 1989). Prajapati et al. (2016) suggested that the 10% concentration of supplementation of the prescribed dose of fertilizer along with extract of either Kappaphycus alvarezii (K sap) or Gracilaria edulis (G sap). They also suggest a greater tuber yield of 32.88 t/ha and 31.30 t/ha, respectively than power, for both 10% G + RDF and 10% K sap + RDF. In addition, marine extract sprays dramatically increased marketable tuber yields and reduced non-marketable and managed potato tuber yields. Sarhan in 2011 backed related results by stating that the beneficial effects of seaweed extracts on potato growth were favorable and that the overall potato production, qualitative as well as quantitative, increased significantly. According to Abetz and Young (1983), the extract of marine algae could increase potato production. Besides that, essential are the stage of the cultivation in which the algae are sprayed. After planting the marine extract spray reported increased tuber yields, improved nitrogen content, total soluble solids, and protein in potato tubers at 30- and 60-days intervals (Haider et al., 2012).

According to Layek et al. (2017), the output of rice grain increased 18.0% with the application of either Kappaphycus (K) or Gracilaria (G). It is suggested that using either 10% of the K or G sappy spray concentrations and the 100% required fertilizer dose is a feasible choice for better growth and grain quality of rice. Patel et al. (2015) and Singh et al. (2015) also record identical rice performance. There are also records of the beneficial effects of marine extract on wheat growth and production (Triticum aestivum).

Wheat grain yield improved dramatically by 19.74% and 13.16% with the application of 7.5% and concentrations of K at 5.0% of K. alvarezii and G. edulis sap, respectively. The same results with the application of K. alvarezzi have also been recorded for wheat (Zodape et al., 2009) (Figure 5.5).

Figure 5.5. Women collecting the raw materials for production of seaweed-based fertilizer.

Sources: Image by Wikimedia.

5.12. CONCLUSION

Agricultural crop performance can be enhanced by the proper use of seaweeds and their derivatives. Different physiological and biochemical processes are involved in plants. As a result of the changes in the rhizosphere populations of plants, the observed effects of foliar or field may be direct or indirect.

Only by using techniques developed for molecular biology, metabolomics, and genomics, are the mechanisms of action of these diverse seaweed products started to unravel now.

The biological effects observed are due not only to small organic molecules, but to polymers in products that can regulate genetic action for systems of plant resistance. The proof at hand shows that the advantages of the use of seaweed products can be undeniable in agriculture. There are fascinating developments yet to be made and it is sure others would be predictable.

REFERENCES

1. (2015). Sargassum Seaweed and its Use in Crop and Livestock Production. [eBook] p. 2. Available at: http://www.cardi.org/wp-content/uploads/downloads/2015/11/Sargassum-Seaweed-and-its-use-in-crop-and-livestock-production-CARDI-Policy-Brief.pdf (accessed on 10 July 2020).
2. Begum, M., Bordoloi, B., Singha, D., & Ojha, N., (2018). Role of seaweed extract on growth, yield, and quality of some agricultural crops: A review. Agricultural Reviews, (of).
3. Craigie, J., (2010). Seaweed extracts stimuli in plant science and agriculture. Journal of Applied Phycology, 23(3), 371–393.
4. Karthick, N., (2014). Biopotentials of Organic Fertilizers (p. 2). [eBook] Shodhganga. Available at: https://shodhganga.inflibnet.ac.in/bitstream/10603/39322/7/07_chapter%201.pdf (accessed on 10 July 2020).
5. Nabti, E., Jha, B., & Hartmann, A., (2016). Impact of seaweeds on agricultural crop production as biofertilizer. International Journal of Environmental Science and Technology, 14(5), 1119–1134.
6. Popescu, M., (2013). Agricultural Uses of Seaweeds Extracts (pdf). [online] ResearchGate. Available at: https://www.researchgate.net/publication/268442947_AGRICULTURAL_USES_OF_SEAWEEDS_EXTRACTS (accessed on 10 July 2020).
7. Suresh, S. P., Kumar, V., Gauri, S., & Harkulkar, G., (2020). Medicinal Value of Seaweeds and its Applications: A Review. [online] ResearchGate. Available at: https://www.researchgate.net/publication/280527895_medicinal_value_of_seaweeds_and_its_applications_-_a_review (accessed on 10 July 2020).

CHAPTER 6

Medicinal and Cosmetic Application of Seaweeds

CONTENTS

In this chapter, the medicinal and cosmetic application of seaweed has been discuses, in addition to the various medicinal properties like antioxidant property, antifouling property, and antimicrobial properties.

The biological properties of brown seaweed in context of its cosmetic application and the seaweed beauty enhancer, protect skin against harmful UV rays, and have anti-inflammatory and anti-allergic properties too.

6.1. INTRODUCTION

One of the ocean's essential living resources is the marine alga, generally referred to as seaweed. Agar, carrageen, and alginate are common sources of seaweed used to feed humans, animal feed, plant fertilizers, and a producer of variety of chemicals.

The seaweed has also recently gained strength as modern biological science experimental systems and advanced aquaculture systems. In our everyday lives, aquatic items are used in one form or the other.

In the manufacturing of toothpastes, soaps, shampoos, cosmetics, milk, ice creams, beef, cooking fruit, air refreshers and many other things certain seaweed polysaccharides are used, for instance. Seaweeds constitute a core of the diet in many Eastern countries such as Japan, China, Korea, etc.

Below are some classic examples of marine applications. In the paper manufacturing agar is commonly used, and also in culture media, packaging material, photography, clothing, plywood manufacturing, food processing, meat, cosmetics, and the pharmaceutical industries. In the food industry, carrageenan is widely used.

It has been well-known for its value in the manufacture of sausages, beef, balls of meat, ham, poultry, and fish preparations, chocolates, dessert gels, iced creams, juice, marmalade, and sardine sauces. In the production of non-food items such as alcohol, air fresheners, textiles, toothpastes, hair shampoos, tissues, culture media, fungicides it is also used.

The use of alginate is known in frozen products, fillings of pastries, syrups, bakeries, dry mixes, meringues, frozen cakes, chocolate instantaneously, fried fillings, artificial food, patisserie, cake gels, made meats, dressings for salads, meat, and flavor sauces. In coastal areas in folk medicine has used a wide variety of species of seaweed.

The coastal populations have recognized the tonic and curing powers of algae from generation to generation, considering the exceptional mineral content of seaweed and are the base of much of their healing properties. The contemporary diets of seaweeds have been used in all near the coast societies such as Japan and China, Hawaii, Wales, Scotland, Iceland, and Ireland.

The explanations for the success of some of these treatments are now clarified with advances in medicine, as scientists at present are more advanced and know the various nutritional and biochemical properties of varying algae or seaweed. The seaweeds contain various medicinal properties. Seaweeds are high in iodine like *Taxiformis* spp., *Asparagopsis* spp. and *Sarconema* spp. could also be used to monitor goiter disease caused by thyroid gland expansion.

The necessary amino acids that are not present in other vegetable food products in the human diet are in the marine algae. The most significant items like agar-agar and algin are produced by seaweed. The algae cell walls comprise tile that is colloidal carbohydrates. Agar-Agar is derived from marine algae, including *Gelidiella acerosa, Gracilaria edulis, G. folifera, G. corticata,* and *G. verrucosa*.

Chennubhotla (1996) extracts algine from species of *Sargassum* and *Turbinaria*. Seaweeds are a source of non-phytoplankton production; they provide feed for associated grazers and make a major contribution to the benthic detritus food chain.

In an ecological sense, seaweeds give the structural integrity of other biotopes, in particular on low-energy shores, where they are primarily used in size and space occupants. Fish and invertebrates also have significant ecosystems in seabed grounds.

They can be used to study diversity patterns and are particularly useful in the planning and sustainable use of the sea on the coast (John and Lawson, 1991; Jennings et al., 2001; Bolton et al., 2003). They are also useful as climate change indicators. An oldest evidence of seaweed use in a late Pleistocene settlement in Chile was 13,000 years ago. Further archaeological evidence indicates that algae have been used in folk medicine in Japan (13,000–300 BC), China (2700 BC), Egypt (1550 BC), India (300 BC) for several thousand years (NAAS, 2003; Teas, 2005). Many of the algae are used in the Asian and Pacific regions as food as outstanding sources of antioxidants, amino acids, carbohydrates, fats, lipids, growth hormones, micro, and macro elements including iodine.

Furthermore, the presence of goiter in Japan and other South-East Asian populations is very uncommon (King, 2007). The commercial usage of marine algae originated from soda and potash of brown algae for the manufacture of soap, glass, and iodine (Dhargalkar and Verlecar, 2009) while the industrialization of marine polysaccharides or phycocolloids is known as thickening.

The production of the phyco-supplementary industry is rapidly increasing, with various applications for medicinal drugs, botanicals, nutraceuticals, cosmeceuticals, additives in fish and animal feed, cultivation, and soil additives, natural pigment sources, bioactive substances, antivirals, etc., (Chopin, 2007).

The commercially available marine macroalgae varieties are generally known as algae. Macroalgae may be classified according to their nutrient and chemical composition, as red algae (Rhodophyta), brown algae (phaeophyta), or green algae (Chlorophyta).

The seaweeds are a major source of natural bioactive substances. Any of the essential components of medicinal interest contain antibiotics, antioxidants (Lekameera et al., 2007), anticoagulants, anti-ulcer agents, and radiation suspenders.

Fresh and dry algae are widely eaten by people who live in coastal areas in particular. It is stated in literature that a substantial amount of protein, vitamins, and minerals important to human nourishment are found in the edible algae (Fayaz et al., 2005).

Approximately 35% of freshly found chemicals were marine algae, accompanied by sponges (27%) and cnidaris (22%) between 1977 and 1987 (Smit, 2004). At present, Carraguard, the carrageenic, non-spermicidal form of red algae, is a promising drug able to avoid the spread of HIV/AIDS and the women's chance of contracting the disease (Robertson and Andersson, 2007).

The regular dietary seaweed intake in Japan and Korea is attributed to low HIV or AIDS prevalence and low incidence of different cancers (Teas, 2005). However, as demand for phycocolloids is projected to increase per year by 8–10% (Dhargalkar and Verlecar, 2009), a substantially higher demand for natural seaweed products should be anticipated as new applications of phyco supplements culture and the usage as food are growing quickly.

6.1.1. Cosmetics

Seaweed extract is always found on the cosmetic packaging list of ingredients, especially in facial, hand, and body creams or lotions. This applies generally

to the use of alginate or carrageen in the product and the uses of cosmetics have previously been addressed.

In place of derivatives from cosmetics, the use of algae is quite limited. Mashed algae, packed in sachets, were often sold in addition to essential oils as a bath water additive. Seaweed meal along with bath salts has also been expected to sell.

In some last years, thalassotherapy, particularly in France, has come into fashion. A variety of therapies are carried out including hydrotherapy, massages, and various marine mud and algae therapies, using mineral-rich seawater. One treatment is to cover the body with fine particles of seaweed, wrap it in clinging wrap, and warm it with the infrared lamps.

In this treatment, a person is treated. It is thought to be useful in different ways, such as rheumatic pain relief or cellulite extraction. In massage creams, there is also utilization of paste mixtures that aim to return elasticity and strength to the skin quickly.

The paste of seaweed is made by freezing or crushing. The algae are washed, cleaned, and then frozen then grinded. They are pressed against or crushed against a grinding wheel, and sometimes freeze with liquid nitrogen to make the frosted material sprinkler and easier to grind or crush. The effect is a fine green seaweed paste.

6.2. MEDICINAL PROPERTIES OF SEAWEED

6.2.1. Antioxidant Properties

The quest for natural antioxidant compounds has attracted tremendous interest over the last decade, and the number of antioxidant and oxidative stress publications has almost quadrupled (Huang et al., 2005).

In diseases like chronic inflammation, atherosclerosis, cancer, and cardiovascular disorders and aging methods there is a greatly beneficial effect of antioxidant compounds (Kohen and Nyska, 2002), and in manufacturing of medicine, production of food and in the cosmetic industry there can be huge application on a commercial level.

In fact, the belief in using antioxidants from natural sources is greatly heightened by customer demand for natural foods and anxiety over the

possible harmful effects of conventional antioxidants (Safer and al-Nughamish, 1999). The marine algae are exposed to light and oxygen, like other photosynthesizing plants, that result in the generation of free radicals as well as other strong oxidizing agents are caused by getting both the light and oxygen like plants photosynthetic process in the marine algae.

The loss of oxidative degradation and stabilization during storage in the structural components of macroalgae, i.e., polyunsaturated fatty acids (PUFAs), indicates that their cells have antioxidants safety mechanisms (Fujimoto, 1990; Matsukawa, 1997).

In addition, algae have defensive enzymes (superoxide dismutase (SOD), peroxidase, glutathione reductase, catalase (CAT)) and antioxidant molecules (phlorotannins, ascorbic acid, tocopherols, carotenoids, phospholipids, chlorophyll based compounds, bromophenols, catechins, amino acids, polysaccharides, etc.), close to those present in vascular plants (Fujimoto, 1990; Le Tutour et al., 1990).

Several regional regions have studied the antioxidant effects of seaweed extracts, but only a few experiments have been conducted on tropical seaweed species (Anggadiredja et al., 1997; Lim et al., 2002; Fallarero et al., 2003; Santoso et al., 2004). Tropical macroalgae's antioxidant activity is surprising given that these species are expected to develop an amazingly effective antioxidant defense system due to the high UV radiation in the tropical environment.

In addition, prior studies have shown that UV radiation promotes antioxidant protection in macroalgae (Aguilera et al., 2002; Bischof et al., 2002). There are several populations of macroalgae which are subject to heavy solar irradiation in the Gulf of Mexico and the Caribbean coast of Yucatan and Quintana Roo and some of them have never been researched for their antioxidant activities.

The present research measured the antioxidant ability of 48 species of macroalgae from the coasts of Yucatan and Quintana Roo (Mexico) by evaluating the radical scavenging capacity of 2,2-diphenyl-1-picrylhydrazyl (DPPH) and the total phenolic compound content in alcoholic extracts.

The number of reactions and mechanisms has been linked to antioxidant activities: chain initiation avoidance, transition metal ion catalyst binding, reduction efficiency, radical scavenging, etc., (Frankel and Meyer, 2000; Huang et al., 2005).

Thus, the basic antioxidant actions of the most active extracts were also characterized in order to better understand the antioxidant mechanisms involved, using various biochemical methods: reduction of activity, superoxide anion scavenging activity and inhibition of lipid peroxidation.

The *Turbinaria ornate* seaweed extracts had significant activity against both positive and negative bacteria as opposed to normal ampicillin. From the present analysis, it is clear that methanol derives from *T. ornata* may be used in pharmaceutical industry as a strong source of antimicrobial agent.

However, it is necessary to evaluate the active components responsible for antibacterial activities. Therefore, it is proposed that further work on isolating and defining the antimicrobial components in it could be carried out according to Vijayabaskar and Shiyamala (2011).

Many marine algae compounds demonstrate antibacterial behaviors (Thirumaran et al., 2006; Vairappan et al., 2001). The bioactive efforts have been shown to possess many metabolites isolated from marine algae (Kim and Lee, 2008; Venkateswarlu, 2007; Yang, 2006). Of the numerous chemically classified chemicals, antioxidants are the most commonly studied.

In fact, the essential role of antioxidants in human health has also been shown to raise market interest in these goods and their production. Sea algae function as essential natural bioactive products (Iliopoulere and Metzger, 2002).

From cancer to coronary and degenerative disorders (Iliopoulere and Metzger, 2001; Croke, 2003), oxidative stress is an important element in the origin of the disease. Fayaz et al. (2005) stated the utility of *Kappaphycus alvarezzi* for various medicinal products, including antioxidants for use as a nutraceutical or health food supplement. It is also recognized that various sections of thalli vary in their antimicrobial capacity.

6.2.2. Antifouling Properties

The type of adhesion on all submerged surfaces of microbial slime, macroalgae, and invertebrates, like the barnacles, modules, and sponges is called marine biofouling. This phenomenon has various implications and the consequences of biofouling have been faced by many sectors, for example, aquaculture, and shipping.

For instance, ship-hull marine biofouling increases fuel use, hull drag, and thus cleaning and painting, resulting in the projected costs of

approximately. For the United States Navy alone, 1 billion dollars a year (Callow and Callow, 2002).

Furthermore, the engine activity decreases with the bursting of hulls and propellers, increasing carbon dioxide emissions, and thus adding greatly to global warming. In the process known as epibiotic, marine organisms themselves are affected by biofouling (Wahl, 2008).

Their surface colonization may cause a transition from a benthic nature to a planktonic nature, a shift from a mobile to sessile life or the start of pathogenesis and finally the host's death (Steinberg et al., 2002).

Maritime organisms have developed mechanisms for fouling, including mechanical (molting or slaughter, toughness, physical defenses) or chemical defense (surface pH compounds) (Stoecker, 1978; Davis et al., 1989; Bobzin and Faulkner, 1991; Wahl and Sonnichsen, 1992; Becker and Wahl, 1996; Hellio et al., 2009).

Amongst these organisms, marine algae are responsible for producing a broad range in chemically active metabolite as an aid to prevent or reduce access to light and nutrients from protected themselves from other settling organisms (Henrikson and Pawlik, 1995; Wahl, 2008; Hellio et al., 2009).

The active metabolites are reported to be effective in biofouling prevention from several species of marine macrose and microalgae (Fusetani, 2004; Maréchal et al., 2004; Barbosa et al., 2007; Tsoukatou et al., 2007; Cassano et al., 2008; Culioli et al., 2008; da Gama et al., 2008; Mokrini et al., 2008; Plouguerné et al., 2008).

For typical defensive brawn algae metabolites, terpene, and phlorotannin, this has already been recorded. In 2005, Vallim et al. discover that *Dictyotacean diterpenes* spatial distribution resembles the differentiation of geography and of chemistry. The diterpene composition has also been qualitatively varied in brown seaweed populations *Stypopodium zonale* (Pereira et al., 2004).

The overall content of sphlorotannins in algae was shown to rely on environmental factors such as the abundance of water, light, and nutrients (Amsler and Fairhead, 2006; Jormalainen and Honkanen, 2008).

There seems to be a variation in the overall phlorotannin level in algae populations at local level (Pavia et al., 1999; Hemmi and Jormalainen, 2004; Koivikko et al., 2008). The genetic distinction between populations and/or reactions to local environmental changes may explain such variation (Koivikko et al., 2008).

6.2.3. Antimicrobial Properties

The seaweeds can produce a large range of secondary metabolites that are represented by a wide array of bioactive activities and are regarded a cause of bioactive compounds. While seaweed grows in a rough climate, photodynamic damages during the metabolism are uncommon. This suggests the presence of antioxidant compounds and processes on seaweed cells (Matsukawa et al., 1997).

Seaweeds are an important source of antimicrobial compounds, including omega-3 fatty acids, antioxidants, and other bioactive compounds and therefore are an important source of nutraceutical products and functional foodstuffs (Yuan, 2008).

The antioxidant-antimicrobial compounds of brown, red, and green algae were identified, such as polyphenols, flavonoid, and polysaccharides (Cox et al., 2009; Zaragoza et al., 2008). Brown algae were shown to be active against a variety of gram-positive and gram-negative species by Ara and others in 2002. According to Nagayama et al. (2002) the bactericidal activities of *Ecklonia kurome* phlorotannins, brown-algal phenolic compounds like eckol and related eckol.

The variety of polyphenolic compounds like the catechins, flavonol, and flavonol glycosides have been reported and found to produce antioxidants and antimicrobials in red and brown algae extracts of methanol (Hosokawa et al., 2006).

The brown algae *Sargassum sagamianum* has antibacterial properties in the extracted sargaquinoic acid derivatives (Horie et al., 2008). The impact on *Listeria* spp., *Salmonella abony*, *Enterococcus faecali,* and *Pseudomonas aeruginosa* of various doses of brown seaweed (*Himanthalia elongata*).

The different observations of the anti-microbial activity of seaweed extract on different species have been available (Taskin et al., 2010; Cox et al., 2009; Ely et al., 2004; Nagayama et al., 2002) but there are no studies of the growth inhibition modeling accessible. *In vitro* antibacterial activities against gram-positive and gram-negative bacteria have been demonstrated in extracts and active components of different algae.

The antimicrobial activity development was considered to imply the seaweed's capacity to synthesize secondary bioactive metabolites (Del Val et al., 2001).

There are various macroalgae compounds with a number of biological activity studies, including antivirals (Richards et al., 1978), anti-tumors

(Espeche et al., 1984), anticoagulants (Athukorala et al., 2006), and anti-fouling (Marechal et al., 2004). Black, brown, and red algae have been shown to have compounds of cytostatic, antiviral, anthelmintic, antibacterial, and antifungal activity (Newman et al., 2003).

Furthermore, given their high taxonomy diversity, research into new biologically active algae compounds is considered a quasi-unlimited field. Antioxidants are the most commonly studied of the various compounds with functional properties.

In addition, the important role of antioxidants in human health has been shown to increase consumer interest and demand for such products. A major resource for bioactive natural products is marine algae (Metzger et al., 2002). The phenolic substances were discovered in Brazilian red algae.

The nutritional and antioxidant property of *Kappaphycus alvarezzi* and its antimicrobial potential differs in various parts of the thalli (Fayaz et al., 2005). Similarly, some pharmacologically active compounds are found or released by micro-algae.

The dinoflagellates, for example, *Gymnodinium* sp. and the *Gonyaulax* sp. produces alkyl guanidine compound that impact the central nervous system. The pharmacological activity is displayed by Rivularia's brominated biindoles. *Gracilaria lichenoides*, red algae, excretes prostaglandin and the compound C20 (Masuda et al., 2005).

The seaweeds are known to include reactive antioxidant molecules such as ascorbates and glutathiones (GSH) as fresh as well as secondal micro-metabolites, which include the carotenoids (α, β-carotene, fucoxanthin, astaxanthin), mycosporins such as amino acids (mycosporine-glycin), and catechins (e.g., catechin, epigallocatechin). Phlorotannins work as antioxidants and antibacterial compounds in Brown algal polyphenols (Kuda et al., 2007) (Table 6.1).

Table 6.1. *List of Some Reported Algae for Antimicrobial Activity*

Number	Name	Microorganisms	References
1.	*Ceramium indica, Enteromorpha Intestinalis, Ulva lactuca, Gelidiella acerosa, Corallina officinalis, Gracilaria corticata, Sargassum sp., Galaxaura oblongata, Grateloupia sp., Ceramium indica, Laurentia pedicullata, Padina gymnospora, Stoechospermum marginatum, Cystoseira sp., Caulerpa racemosa, Caulerpa scalpelliforrmis, Spathoglossum variabile, Halimeda tuna, Gelidiopsis gracilis, Iyengaria stellata, Gelidium sp., Udotea indica, Gastroclonium iyengarii, Heterosiphonia sp.*	*B. cereus* ATCC11778, *M. flavus* ATCC10240, *K. pneumoniae* NCIM 2719, *P. testosteroni* NCIM5098, *C. freundii* ATCC10787	Nair et al., 2007
2.	*Corallina officinalis, Cystoseira barbata, Dictyota dichotoma, Halopteris filicina, Cladostephus spongiosus f. verticillatus, Ulva rigida*	*S. aureus, M. luteus, E. faecalis, E. coli, E. aerogenes* O157:H7	Taskin et al., 2007
3.	*Gracilaria tikvahiae, Ulva lactuca, Ulva fasciata, Sargassum fluitans*	*E. coli, S. aureus, S. enteriditis, S. epidermidis, P. aeruginosa, S. faecalis, C. albicans*	Oranday et al., 2004
4.	*Ulva lactuca, Padina gymnospora, Sargassum wightii, Gracilaria edulis*	*S. aureus, V. cholerae, S. dysentriae S. bodii, S. paratyphi, P. aeruginosa, K. pneumoniae*	Vallinayagam et al., 2009
5.	*Trichodesmium erythraeum*	*B. subtilis* MTCC441, *E. faecalis* ATCC29212, *S. aureus* ATCC 25923, *E. coli* ATCC25922, *K. pneumonia* ATCC15380, *P. vulgaris* MTCC1771, *P. aeruginosa* ATCC27853, *E. amylowora* MTCC 2760, *S. typhimurium, T. rubrum* 57/01, *T. mentagrophytes* 66/01, *T. simii* 110/02, *Scopulariopsis* sp. 101/01, *A. niger* MTCC1344, *A. flavus, B. cinerea, C. albicans* MTCC227	Thillairajasekar et al., 2009
6.	*Sargassum sp.*	*B. subtilis* CCR 12, *E. coli* MTCC 443, *S. aureus* MTCC 96	Patra et al., 2008
7.	*Eucheuma denticulatum*	*S. aureus, S. Pyogenes*	Al-Haj et al., 2009

8.	*Ulva rigida, Enteromorpha linza, Padina pavonica, Colpomenia sniosa, Dictyota linearis, Dictyopteris membranacea, Cystoseira mediterranea, Ectocarpus siliculosus, Ceramium rubrum, Gracilaria gracilis, Acanthophora nojadiformis*	*Candida sp., E. faecalis, S. aureus, S. epidermidis, P. aeruginosa, E. coli*	Tuney et al., 2006
9.	*Kappaphycus sp., Padina sp.*	*P. fluorescence, S. aureus, V. cholera, P. mirabilis, B. megaterium, Citroacitro sp., Entrobacter sp., K. pneumoniae, E. coli, S. typhimurium, S. fexneri, P. vulgaris, P. aeruginosa, V. cholera*	Rajasulochana et al., 2009
10.	*Sargassum dentifolium, Laurencia papillosa, Jania corniculata*	*B. subtilis, S. albus, S. faecalis, E. coli, C. albicans, A. flavus*	Shanab et al., 2007
11.	*Colpomenia sinuosa, Dictyota dichotoma var. implexa, Petalonia fascia, Scytosiphon lomentaria*	*B. subtilis* ATCC6633, *S. aureus* ATCC6538P, *E. aerogenes* ATCC 13048, *E. coli* ATCC29998, *P. vulgaris* ATCC6897, *S. typhimurium* CCM 583, *methicillin-oxacillin-resistant S. aureus* ATCC43300, *E. coli* (O157: H7) RSSK 232, *C. albicans* ATCC10239	Demirel et al., 2009
12.	*Synechocystis sp., Himanthalia elongata*	*S. aureus* ATCC25923, *E. coli* ATCC11775, *C. albicans* ATCC 60193, *A. niger* ATCC16404	Plaza et al., 2010
13.	*Microchaete tenera, Nitella tenuissima, Sphaeroplea annulinai*	*P. vulgaris, B. cereus, E. coli, P. aeruginosa, A. niger, A. flavus, R. nigricans*	Prashantkumar et al., 2006
14.	*Ulva fasciata, Caulerpa cupressoides*	*B. subtilis, S. epidermidis, S. aureus, C. freundii, E.coli, E. aerogenes, K. pneumoniae, M. morganii, P. aeruginosa, S. typhi, S. typhimurium, S. enteritidis, S. cholerae-suis, S. marcescens, V. cholerae*	Lima-Filho et al., 2002

15.	*S. marginatum, St. marginatum, P. tetrastromatica, P. gymnospora, C. implexa, D. australis, D. bartayresiana, S. aspermum, Asparagopsis taxiformis, Amphiroa fragilissima, Jania reubens, Caulerpa racemosa, Caulerpa peltata, Caulerpa taxifolia, Codium fragile, Chlorodesmis fastigiata*	*B. subtilis, E. coli, Pseudomonas sp., S. pyrogenes, S. aureus, P. vulgaris, K. pneumoniae, S. morganii, C. albicans*	Kotnala et al., 2009
16.	*Gracilaria corticata, Ulva fasciata, Enteromorpha compressa*	*V. alginolyticus, P. aeruginosa, A. hydrophila, E. tarda, P. fluorescens, A. hydrophila*	Choudhury et al., 2005
17.	*Valonia aegrophila, Ulva Halimeda opumtia, Halimeda tuna, aulerpa racemosa, Caulerpa Mexicana*	*S. aureus, B. subtilis, E. coli, C. albicans*	Mtolera et al., 1996
18.	*Cystoseira tamariscifolia*	*S. cerevisiae, C. albicans, Debaryomyces sp., Kluyveromyces sp strains), A. flavus, Penicillium sp.*	Zineb et al., 2004
19.	*Spirulina platensis*	*S. faecalis, B. subtilis, S. aureus, S. epidermidis, P. aeruginosa, E. aerogenes, E. cloacae, E. coli, S. typhimurium, P. vulgaris, C. albicans*	Ozdemir et al., 2004
20.	*Dictyopteris membranaceae, Cystoseira barbata*	*S. faecalis, B. subtilis, S. aureus, S. epidermidis, P. aeruginosa, E. cloacae, E. coli, S. typhimurium, C. albicans*	Ozdemir et al., 2006
21.	*Enteromorpha linza*	*S. aureus, S. epidermidis, S. fecalis, B. subtilis, S. typhimurium, P. aeruginosa, E. cloacae, E. coli, C. albicans*	Sukatar et al., 2006
22.	*Jania rubens*	*S. aureus, S. epidermidis, S. faecalis, B. cereus, B. subtilis, S. typhimurium, P. aeruginosa, E. cloacae, E. coli, C. albicans*	Karabay-Yavasoglu et al., 2007
23.	*Polysiphonia virgata*	*M. smegmatis, M. tuberculosis, multidrug resistant M. tuberculosis*	Saravanakumar et al., 2008
24.	*Ulva lactua, Halimeda gracilis, Gracilaria edulis, Hypnea musiformis, Turbinaria conoides, Sargassum myricystum*	*E. coli, P. aeruginosa, S. aureus, K. pneumoniae, E. faecalis*	Kolanjinathan et al., 2009

25.	*Gracilaria changii*	*S. aureus, P. aeruginosa, P. mirabilis, E. coli, A. calcoaceticus, S. saprophyticus, S. marcescens, K. pneumoniae, A. nitratus, S. paratyphyi B, B. licheniformis, Micrococcus sp., S. pidermidis, C. preundii, E. enterocolitica, S. typhi, B. pseudomallei, B. subtilis, Erwinia sp., B. cereus, E. aerogenes, C. albican, R. rubra, C. neoformans, S. cerevisiae, T. viride, Rhizopus sp., Mucor sp., Penicillium sp., Fusarium sp., T. mentagrophytes, F. oxysporium, A. niger, A. flavus*	Sasidharan et al., 2009
26.	*Ulva fasciata*	*V. parahaemolyticus, V. alginolyticus, V. vulnificus*	Chakraborty et al., 2010
27.	*Dictyota acutiloba*	*Methicillin resistant S. aureus, methicillin susceptible S. aureus, Enterobacter sp., P. aeruginosa, S. typhi, B. subtilis, K. pneumoniae, C. albicans, A. niger*	Solomon et al., 2008

Sources: Seaweeds-A novel, untapped source of drugs from the sea to combat Infectious diseases by S. Chanda, Dave, R., Kaneria, M., and Nagani, K.

6.3. COSMECEUTICAL APPLICATIONS OF BROWN SEAWEEDS (FIGURE 6.1)

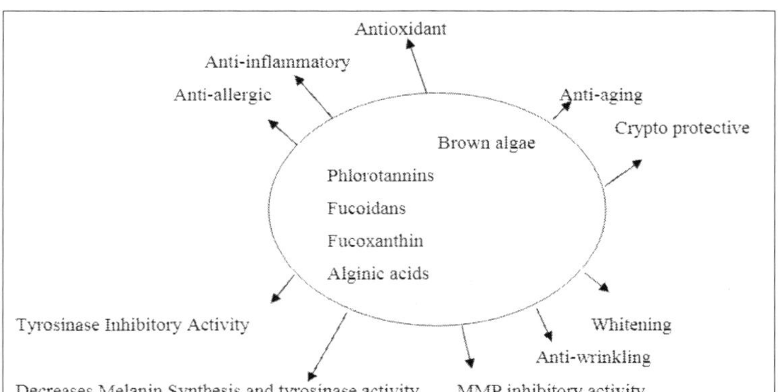

Figure 6.1. Properties of brown algae.

6.3.1. Antioxidants

The cells are protected against the reactive oxygen compounds and this protection is provided by the compounds name as Antioxidant. This reactive oxygen basically leads to a damaging effect on the cells. Some of the examples of these reactive oxygen compounds are namely such as single oxygen, superoxide, peroxyl radicals, and hydroxyl radicals. The problem of cellular damage basically occurs when there is entirely an imbalance created between the ROS and the antioxidants.

Due to these cellular damages many serious problems are caused within the body in simple language many serious diseases can be caused, and these are diseases are linked to cancer, inflammation, and aging. Cancer is one of the deadliest diseases and the one which is difficult to cure. Further, it is observed the skin is protected against the damaging effects by the means of antioxidants (Weber et al., 1997).

Furthermore, they are considered as one with great advantages and due to their effects on the cells, they have been popularized throughout the world. Also, nowadays these antioxidants are used in various products and all because of their excellent functions. They are used as the primary ingredients in cosmetics as they protect the skin cells from damaging when exposed to harmful substances and they are also used as functional ingredients (Lupo, 2001).

Many natural antioxidants sources can be nowadays used or utilized in many industrial applications. And in the current scenario, various researchers are paying attention towards these natural antioxidant sources so that these sources can be optimally utilized. Moreover, within the cosmeceuticals applications, these natural antioxidants are attracted and the interest towards them has been increased (Sun and Ho, 2005).

The phloroglucinol, eckol, and dieckol are basically three phlorotannins which are purified from the brown seaweeds. These phlorotannins are basically found to perform the antioxidant activities (Ahn et al., 2007).

To determine the protecting effects against in H_2O_2-mediated DNA damage there is a technique which basically in the study, measured by the electron spins resonance (ESR) spectrometry technique. And this is used for comet assay and for the scavenging effects of free radicals and these two was only the ones used in this determination. Further, by enhancing the cellular antioxidant activity and modulating cellular signal pathway were basically, by which eckol protects the V79-4 cells against the oxidative damage (Kang et al., 2005).

With the means of the novel phlorotannin, derivatives apart from the promising antioxidant properties of *E. cava* were also reported by it (Li et al., 2009). Moreover, among the isolated compounds, there are 60 bieckol, dieckol, and flucodiphloroethol and by the contrast, they have also reported the profound antioxidant properties of these compounds. The V79-4 cells are basically protected against the oxidative stress induced cell through up regulation of HO-1 only by the means of the triphloethol as a phlorotannin and this reported by a recent study (Kang et al., 2007).

By reducing the ROS effect, the triphenol-A basically protects the cells against the oxidative damage induced by radiation and this radiation is the one which is induced by reducing the ROS effect (Kang et al., 2006). Some of the compounds were found similar to those of the antioxidant activity with L-ascorbic acid and these compounds are an antioxidant phloroglucinol which was isolated and isolated compound, and this was also been understood by another study (Lee et al., 1996). Moreover, the antioxidant effects of the fermented *E. cava* processing by-product and this were demonstrated by the means of one of the previous works (Wejesinghe et al., 2011).

By the means of phlorotannins it was later reported that the antioxidant activity of *E. cava* processing by-product mainly by it. A natural carotenoid which is isolated from the brown seaweed *S. siliquastrum* (fucoxanthin) and this was revealed by the experimental findings that cytoprotective effect of fucoxanthin (Heo et al., 2008). By the means of the nuclear staining with Hoechst dye a protected effect against the oxidative stress, induced cell apoptosis was also demonstrated. Further, it was also observed that the prominent antioxidant was basically present within the fucoxanthin which was isolated from the *S. siliquastrum* and it was clearly indicated by the recent findings.

Within the three Indian brown seaweeds which are namely *S. marginatum, P. tetrastromatica* and *T. conoides* basically *in vitro* possess the various antioxidant activities and this was mainly reported (Chandini et al., 2008). When compared to the other two species it has been observed that *Turbinaria conoides* significant higher phenol content.

Furthermore, a dose dependent manner was observed in the case of the *in vitro* antioxidant activity of methanolic extracts from all the three seaweeds. Also, by the means of another recent study it has been observed that the methanolic extracts of brown seaweeds *in vitro* and cellular antioxidant were reported (O' Sullivan et al., 2011).

The most effective extracts in terms of antioxidant activity were basically reported by the brown seaweeds *F. vesiculosus* and *F. serratus* extracts which has been observed. Further, it was also observed or anticipated that in the cosmeceutical preparations there can be an effective use of the brown seaweeds. Moreover, one of the principal mechanisms of aging is known as the free radical damage.

From the harmful rays of the sun that are usually termed as the ultraviolent rays or from other harmful pollutants, the skin can basically suffer the additional skin damage. Further to break down the collagen and the other vital skin function can be mainly caused by the means of the free radicals basically when it comes to the wrinkles.

As the treatment with some antioxidants, such as ascorbic acid there cans effective uses, which are the tropical antioxidants basically provides some protection against the environmental damage to the skin. Also, it has been observed that this antioxidant can effectively slow down the skin aging all because of the above-mentioned reasons or the factors.

Further to prevent the skin aging and also to be effective enough to basically enhance the resistance towards the oxidative stress there are tocopherols and polyphenols (Masaki, 2010). All because of the antioxgenation properties of the cosmeceuticals they can not only improve the appearance of the skin but also can benefit the health of the skin and also these are easily commercially available products (Guan et al., 2005).

To interrupt the free radical damage the properties which an ideal skincare product should bear is that it should basically contain a potent assortment of stable antioxidants to interrupt the free radical damage. And this ideal product can be present in any form either in a liquid gel, serum, lotion, or cream. To help the skin to repair systems one of the most effective ingredients of the cosmecueticals us the antioxidants which can in turn lead to the interruption of the radical chain processes.

The sustained research can only reveal the good natural sources of the antioxidants can be basically considered to the brown seaweeds. And also due to the free radical damage is one of the major causes of the skin aging therefore in contrast the antioxidant effects of phlorotannins ideally coupled with cosmeceuticals.

6.3.2. Protective Effect against UV Light-Induced Skin Damage

The major environmental risk factor for all the skin cancers is the ultraviolent rays of the sun which are basically the most biologically active components of the sunlight (Xu et al., 2000). The different types of skin changes and causing the premature skin aging in addition to the above the sunlight possess these profound effects on the skin.

The serious harm to human and animal health effects this wide range of the possible harmful effects that is basically caused by the different sources of UV radiation. They are also solely responsible for the primarily damage which are related to the skin, eyes, and the immune system. There is a direct interaction with the cellular targets such as the DNA and this inferred by the means of evidence (de la Coba et al., 2009).

The precautionary behavior such as the use of sunscreen, sunglasses, or protective clothing basically helps an individual to protect themselves against the harmful UV radiations. Also, there are many factors which are responsible for the human exposure to UV radiations depends and these timing of the outdoor activities and the duration.

The health effects of an individual can be easily influencing, and occurrence of these diseases can due to an individual's skin color and the age of the person. The prevention of UV induces sunburn, photoaging, and the skin cancer could be easily advocated by the means of the tropical application of the sunscreens and they are widely use throughout the world (Torrres et al., 2006).

Further, against the UV damage the protective effects of the active components from the brown seaweeds are therefore discussed by an entire report. The ability to protect against the oxidative stress induced by UV-B radiation is basically consisted by the Fucoxanthin which is a natural carotenoid which is mainly isolated by the marine brown alga *S. siliquastrum*. And these can also be applied as the antioxidant and the cosmeceutical industries (Heo and Heon, 2009).

The cell injury in human fibroblasts could be induced when in contrast it was investigated and confirmed the protective effect of the fucoxanthin on UV-B. In protection mechanisms against the UV radiation damage it was observed that in addition with the fucoxanthin significantly scavenged the ROS by the exposure to the UV radiation; mainly in this protection mechanism, the major role was played by the polyphenolic compounds of brown seaweeds (Connan et al., 2007).

A prominent protective effect which was basically against the photo oxidative stress which was induced by the UV-B radiation was mainly by the diekol isolated from a brown alga *E. cava* (Heo et al., 2009; Ko et al., 2009). It consists of six phloroglucinol units (hexamer) and functional hydroxyl groups moreover, this dieckol is s phenolic moiety it is a kind of phlorotannin.

According to Heo et al. (2010), diplorethohydroxycarmalol is separated from the brown algae I in another latest experiment in carmalol. Okamurae demonstrated strong protection against UV-B intracellular ROS. The provoking of an acute sunburn reaction is due to overexposure to UV radiation (Thiele et al., 1997).

The skin is the preferred target for oxidative stress since it is permanently exposed to UV radiation, from sunlight to pollutants that oxidize the environment. They are an important aspect of active polyphenolic compounds in protection from UV light-induced skin damage, with phlorotanins. In addition to the scientific evidence supporting its use in cosmeceutical formulations (Figure 6.2).

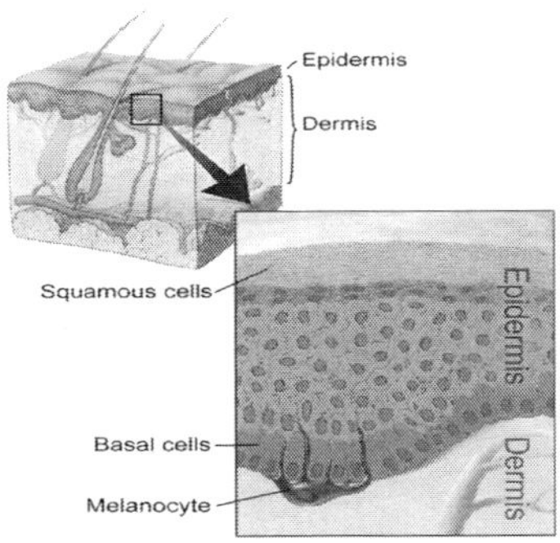

Figure 6.2. Layers of skin.

Sources: Image by Wikimedia.

6.3.3. Tyrosinase Inhibitory Activity and Inhibition of Melanin Synthesis

The main cause of the color of the skin, eyes, and hair is a pigment of melanin. The Melanocytes are cells that contain melanin in the bottom surface of the epidermis of skin and center and the eye middle layer. These cells generate melanin by a process known as melanogenesis.

A rise in melanogenesis is caused by exposure to UV-B radiation as a result in photographic DNA damage (Agar and Young, 2005). Tyrosinase is an enzyme needed to produce the melanin from amino acid tyrosine for melanocytes. Tyrosinase has also been complemented with a sequence of oxidative reactions by the main enzyme in the synthesis of melanin pigment (Gao et al., 2008).

Therefore, the pigmentation of the head, eyes, and hair in mammals is caused by tyrosinase (Likhitwitayawuid, 2008). Due to the considerable industrial and economic impact of this protein's inhibitors, the study of tyro-inhibitory activity has become of importance in recent years (Fails et al., 2009).

In comparison, tyrosinase inhibition may be useful for cosmeceutical depigmentation applications. To date, there have been several inhibitors of tyrosinases, and melanogenesis both naturally and synthetically has been recognized (Sima et al., 2011). The derivatives from marine brown algae *E. cava* and *S. siliquastrum* decreased the *in vitro* and zebrafish animal model the melanin synthesis and tyrosinase production (Cha et al., 2010).

The study found therefore that both brown algal extracts could be utilized in whitening cosmetics as an additive. The phlorotannins separated from *E. cava* have an inhibitor effect, and *E. cava* also has been reported in the B16F10 melanoma cells for mushroom tyrosinase activity, and melanin formation (Yoon et al., 2009). They were evaluating the viability of isolated phlorotannins from *E. cava* as a melanin formation inhibitor and the findings indicate that 7-phloroeckol may prove useful as a novel melanin formation inhibitor in skin creams.

The high inhibitory activity of tyrosinase was observed in *Ecklonia stolonifera* (Laminariaceae) which was part of brown algae (Kang et al., 2004). Their research had shown that derivatives of phloroglucinol, isolated from *E. stolonifera*, through inhibition of tyrosinase production, may be involved in regulation in pigmentation in plants and other animals.

In another experiment, dieckol, and phlorotannin isolated was recommended that *E. cava* is capable of relatively higher inhibition of tyrosinase in comparison the commercial inhibitor of cojic acid tyrosinase (Heo et al., 2009). The findings have showed the potential of specific phlorotannin compound to be used in cosmeceutical applications.

Melanogenesis okamurae have been assessed through an inhibitory action of tyrosinase (Heo et al., 2010). The active diphlorethohydroxycarmalol compound had impressive findings showing higher activity than the commercial whitening agent. It now seems that certain secondary metabolites in brown algae can be attributed with important actions and evidence that the potential whitening effects can be used in cosmeceutical applications.

Furthermore, it generally known that free radicals cause oxidative stress to cause the hyperpigmentation. Consequently, the antioxidant characteristics of brown algae are another advantage to depigmentation (Figure 6.3).

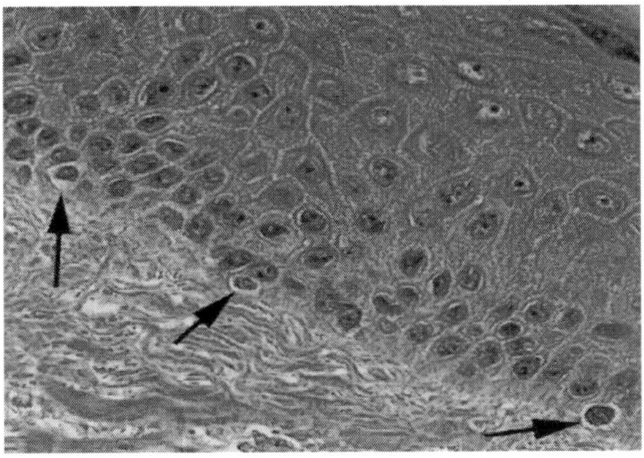

Figure 6.3. Micrograph of melanocytes in epidermis.

Sources: Image by Wikimedia.

6.3.4. Anti-Inflammatories and Anti-Allergies

The natural cell reaction of vascular tissues to harmful stimuli is inflammation, and it is a body protective endeavor to both remove the injurious stimuli and activate a tissue recovery process. Macrophages are essential inflammatory agents (Kazlowska et al., 2010). The activation of NF-kB and AP-1 is influenced by oxidative stress-induced inflammation.

The antioxidant or anti-inflammatory effects of dietary polyphenols have been identified as having unwanted effects of oxidative stress (Han et al., 2007). The use of antiallergic agents may be used for the treatment of allergic reactions and most of these agents act by blocking the release or suppression of the acts of mediators released from their cells. The anti-inflammatory properties of brown marine algae have been supported to date by previous researches.

The powerful effect of E compounds of phlorotannin. The study was based on cava; eckol, 2 dieckol, 6.60-bieckol and 1-(30,50-dihydroxyphenoxy)-7-(200,400,600-trihydroXydroxyphenoxy). The findings of these studies include osteoarthritis, atherosclerosis, and cancer on the host of inflammatory diseases that have a common effect (Ryu et al., 2008).

In addition, these compounds not only displayed the pro-inflammatory cytokines that stimulate expression of human osteoblast MMP-1, -3 and -13, but also revealed regulating activation in non-induced differentiations of ERK and p38 MAP kinases. In addition, E ethanolic extracts are reported by Jung and others in 2009. *E. Cava* inhibited concentration-dependent production of LPS-induced nitric oxide (NO) and prostaglandin E2 (PGE2) and inhibited inducible nitrogen oxide (iNOS) and cyclooxygenase (COX)-2 in BV2.

They reported that as per their results by removing proinflammatory cytokines, E. cava extracts exercise anti-inflammatory effects. In a recent review, dieckol, and 6,60-bieckol, isolated from *E. cava* has been showing notable properties of antiallergy (Le et al., 2009). In addition, they proposed a possible anti-allergic mechanism for eliminated binding activity between IgE and FceRI with the results of the flow cytometric analysis.

There are also more and more new cosmeceutical products in the global market with natural ingredients like the fucoidane and alginate from brown algae. These derivatives are good to be utilized in cosmeceuticals as natural aquatic products (Fitton et al., 2007). The preparations can be made easier in many different orders to manage skin health.

The cosmeceutical industries still are waiting for some serious growth. Cosmeceutical preparation requires careful use of research and use of an appropriate mixture of specific ingredients in the formulation. In the end, the standard cosmeceutical components were required for anti-aging, anti-wrinkling, and whitening to achieve the desired efficiency.

The use of these brown algal extracts or compound products is well linked to the latest trend for compositions of natural products and offers a cosmeceutical solution from basic to advance. Brown seaweeds with their potential cosmeceutical applications representing cosmeceutical candidate ingredients.

6.4. COSMETIC APPLICATION OF SEAWEED

6.4.1. Beauty Enhancer

A surprising similarity between the human skin tissue and the cell structure of algal has contributed to the resolution of many cosmetic and dietary problems. Today, seaweeds in cosmetic products like soaps, shampoos, powders, creams, and sprays have become a basic raw material. In beauty products, France, and Great Britain use algae extensively.

The seaweeds, which moisturize and include amino acids, minerals, and vitamins to nourish your skin, have been found by a skin product manufacturer. Black algae extract such as *Fucus* spp. are applied in a massage therapy which removes impurities from the body, while balancing the pH of the skin, is applied in thalassotherapy.

Few seaweeds have collagen-like molecules, a matter that gives skin elasticity and firmness. Seaweeds and other aquatic goods have been a viable replacement to the chemical-based cosmetics owing to their medicinal properties.

6.4.2. Anti-Wrinkling and Whitening of Skin

The skin is the growing organ of the body and one of the most significant, occupying a region of around two square meters. With its clever defense mechanisms, it is an important obstacle to the outside world (Zhang, 2013).

The dermis and the epidermis are two important layers of the skin. The epidermis is the thin layer of epithelial tissue containing many layers: corneal stratum, translucent stratum, granular stratum, spinoid stratum, and germinative stratum (Brenner and Hearing, 2007; Wollenberg et al., 1996). The layer outside of the epidermis is the stratum corneum.

There are dead cells that are steadily falling with large amounts of keratin. Depending on the stimulus, the thickness of the stratum cornea is determined. The feet soles, for instance, have a bigger layer than the belly's region. Callus is caused by this area of the epidermis being constantly stimulated. There is lucid layer after the stratum corneum.

This layer cannot be found in places where the skin is thinner, not thickened. It contains dead cells or cells that degenerate. There is a granular layer beneath the lucid stratum, a matrix of cells forming granules. The keratin precursors are such structures within the cell.

The spinal stratum that has polyhedral-shaped cells. The cytoplasmic projection, the large number of desmosomes and a material found among cells are mainly glycoproteins and lipoproteins, united these cells. The germinating stratum, the epidermal's innermost layer. Therefore, it is the layer responsible for the formation of the higher levels and therefore the rise of the epidermal epidermis which can be identified as cells under cell division. Bone marrow is home to langerhans cells. They carry huge granules called Birbeck granules and are aplenty dendritic cells in the epidermis.

They are commonly located in lymph nodes and in other organs in histiocytosis. They take part in allergic reactions. They are mobile and dendritic cells that cause skin immune monitoring (Wollenberg et al., 1996). The skin system senses diverse stimuli in the environment and respond adequately to sensing heat with mechanoreceptors, sensation, and thermoreceptor sensors.

These types of receptors can trigger sweat in the sweating bottle, thereby maintaining homeostasis of temperature (Zouboulis, 2004). The subcutaneous layer of the skin where the fat is stored also provides insulation. The skin not only retains oxygen and nutrients, but also other drugs, such as topical steroids.

Further, the skin is rich in 7-dehydrocholesterol and this substance is transformed to vitamin D (cholecalciferol) when exposed to ultraviolet light which can be obtained through the use of dairy products (Denda et al., 2007). The immune function of the skin prevents significant damage because of pigmentation by UV light (Chen et al., 2007).

The epidermis contains dermis that comprises primarily of connective tissue, blood vessels, sweat glands, muscles, fibroblasts, collagen, and elastin. The cross-linked collagen and elastin support the skin. The hyaluronic acid (HA) is one of the main dermal components and is involved in the repair of tissue. In water preservation, HA is of essential significance and can contain water over 1000 times its own size.

The collagen and HA break with aging, however, causes wrinkles and the skin loses firmness. The third layer, the hypodermis, is mainly composed of fat and loose bonding tissue. It provides body insulation, stores energy, and mechanically allows externally applied pressures to be attenuated and dispersed (Wickett et al., 2009) (Figure 6.4).

CHAPTER 7

Extraction of Seaweed: Processes and Challenges

CONTENTS

Seaweeds have assumed great importance in the way that they are being applied across various industries to be used by the humans on several ways. However, the seaweeds have to be extracted from the sea for their further use and the process can be quite cumbersome on some occasion.

This chapter takes the readers through the various challenges that may be faced in the extraction of seaweeds and the ways in which they can be overcome.

7.1. INTRODUCTION

To increase the plant productivity and the production of the food in several regions of the world, the micro and macro algae have long been used and they basically possess various beneficial benefits where they have been applied. It has been mainly observed that the interactions between the soil community and the algae are explicitly complex.

Furthermore, the environmental conditions and the local environment are the factors on which the benefits are depended. The above context relates to the speculation which has been accompanied along with the mechanisms and also to the validity of the results which have been reported.

Mainly for the agricultural use, the first commercial seaweed extracts were done in the 60 years from now mainly. The specific plant organs such as the leaves and the roots were first time treated by these aqueous extracts mainly the direct applications of soluble seaweed were first time constituted this part.

Further, it has been observed that the manorial value or the micronutrients were the two major reasons due to which the earlier concepts that benefits of seaweed and their extracts was. To improve the animal nutrition and the productivity the seaweeds are more likely have been used for the millennia as the fodder supplements.

The specific health benefits, the mechanisms of the action in the animals and the mode of action are the major aspects on the research being conducted mainly focuses on. Further, it has been lately observed that basically the seaweed extracts can modify plants and the animal responses at a fundamental level.

And this is mainly considered only by the means of the improved analytical techniques and the instrumentation which have been coupled along with the use of the molecular genetics.

Basically, to remark ton novel findings and to review key developments which have been done over the years it seems the most appropriate time review all of them. In both the plants and the animal application, a new and exciting vista has been opened for the seaweed extracts.

7.2. CONVENTIONAL PROTEIN EXTRACTION METHODS

In the raw and the unprocessed form in case of the seaweed and the microalgae have poor protein digestibility. Furthermore, in order to improve the bioavailability of the algal protein the matter of concern is explicitly laid on the developing methods for this algal protein extraction and this is the main reason for using this method and developing it.

When compared to the proteins from the other crops to the algal proteins and their methods of extraction are relatively and unfortunately poorly studied topics.

Also, there are some of the particular methods used for the extraction of the algal proteins. Aqueous, acidic, and alkaline methods and followed by several rounds of centrifugation and the recovery are the conventional means used for the extraction of the algal protein. Also, it has been mainly observed that along with the above-mentioned means, they're some of the major techniques used for the extraction of these algal proteins and these techniques are namely chromatography, ultra-filtration, and the precipitation. The above-mentioned techniques are mainly used for the recovery purpose.

Mainly for the extracting proteins namely *A. nodosum*, *Ulva* spp. and *L. digitata* are efficiently extracted by twp particular chemical methods of the extraction and these are namely two-phase acid and the alkali treatments (Figure 7.1).

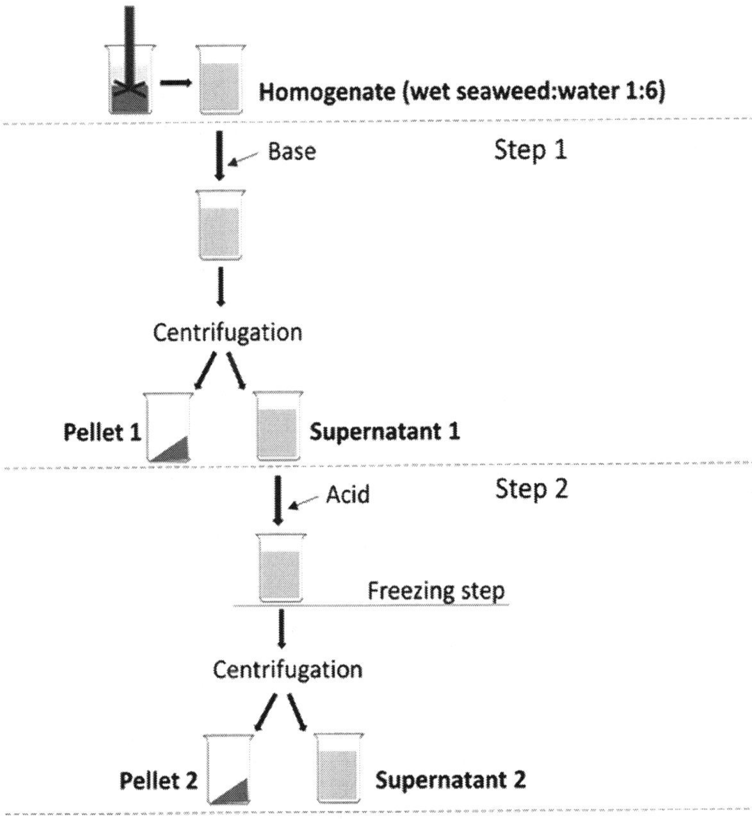

Figure 7.1. Extraction of protein from Swedish red, green, and brown algae.

Source: Image from SpringerLink.

7.2.1. Physical Processes

A significant increase in the extraction yield from the porphyra acanthophora var, acanthophora, sargassum vulgare and ulva fasciata following the immersion in the ultra-pure water is mainly done by the use of a potter homogenizer as it shows significant results along with that of the physical grinding and it has been reported by Barbarino and Lourenço (2005).

Further, to improve the extraction of the algal proteins efficiency alternatively, the osmotic stress could be done and it has been reported. As compared to the high shear force with an Ultra turrax T25 Basic tool (IKA®, Staufen, Germany) (0.74 ± 0.02 g/100 g to the P. palmate (1.02 ± 0.07 g/100

g) a significant higher concentration of water-soluble proteins has been mainly reported by the means of the osmotic shock.

Further, between the two methods it has been basically observed that there was no significant difference in the amount of the total protein which has been extracted. With a concentration of 11.57 ± 0.08 g/100 g *P. palmata*, equating to a yield of 67%, an alternatively, the use of polysaccharides was reported to be a more promising method in case of the protein extraction.

7.2.2. Enzymatic Hydrolysis

Cellulose, galactan, xylans, fucoidan, laminarin, alginates, carrageenans, and floridean starch are the several types of polysaccharides included and basically, seaweed is rich in these several types of polysaccharides.

The protein extraction efficiency was decreased; these polysaccharides can mainly reduce the availability of the algal protein. In order to increase the protein yield enzymes such as polysaccharides can be therefore applied as a cell disruption treatment which is done prior to the protein extraction.

In the protein extraction from the red seaweed species which is namely *C. crispus*, *Gracilaria verrucosa*, and *P. palmata* as a method of combating the tough cell wall basically several polysaccharides (κ-carrageenans, β-agarose, xylanase, cellulase) were used.

7.3. CURRENT PROTEIN EXTRACTION METHODS

Mainly for the commercial use, the protein extraction methods used on algae to date are basically limited. This is due to the concerns of the up-scaling. Further, the extraction of the algal proteins due to the release of proteases from cytosolic vacuoles is mainly affected by the conventional mechanical and enzymatic methods for protein extraction.

It has been observed that these methods are also time consuming as well as laborious. Therefore, there is an explicit need of the improved extraction methods of the cell disruption and the extraction.

Further, the availability if the proteins and the other high value components for later extraction are mainly increased by the means of the pre-treatment with cell disruption techniques which aid the breakdown of the tough algal cell wall.

7.3.1. Ultrasound-Assisted Extraction (UAE)

Ultrasound-assisted extraction (UAE) can be useful for food sources for various applications, including alteration of plant micronutrients to enhance bioavailability, synchronous extraction, and encapsulation, prompting radical sonochemistry to stay away from debasement of bioactive and increasing bioactivity of phenolics and carotenoids by focused hydroxylation (Figure 7.2).

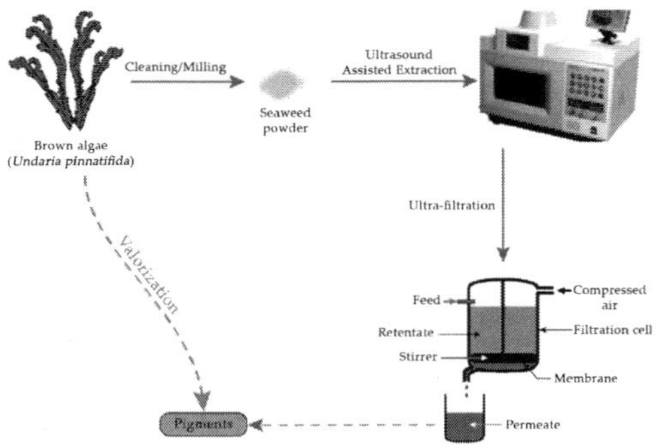

Figure 7.2. *Ultrasound extraction process.*

Source: Image from Science Direct.

The degradative impact of radical sonochemistry, which is the most pertinent perspective regarding improving bioavailability of algal proteins, isn't created by the ultrasound waves, but instead by the formation, development, and implosion of bubbles shaped by what is known as acoustic cavitation.

The violent implosion of these bubbles makes minute districts of outrageous weight and temperature, bringing about the chemical excitation of the sonicated fluid and its substance, encouraging the molecule breakdown and debasement of the particular compound. The significant benefits of UAE are its quick handling time, non-thermal properties, and low solvent utilization, bringing about a finished item of higher value with decreased downstream processing required.

Ultrasound pre-treatment was accounted for to expand protein extraction of *Ascophyllum nodosum* with acid and base treatment alone by 540% and

27%, separately, in addition to diminish handling time from 60 min to 10 min.

This sensational increment in protein yield was proposed to be because of acid hydrolysis alone being inadequate to dissolve the hard cell wall. Ultrasound-supported extraction was likewise assessed in microalgae for various value-added segments, in spite of the fact that there have been moderately hardly any investigations that have concentrated on ultrasound for improved protein extraction.

Keris-Sen and associates (2014) detailed that ultrasound at a power intensity of 0.4 kWh·L−1 yielded the ideal amounts of proteins from wastewater treatment microalgae from the Chlorococcales order of the Chlorophyceae class (e.g., *Scenedesmus* sp.).

Ultrasound treatment of C. vulgaris fundamentally gave rise to crude protein edibility in rodents contrasted with electroporated and untreated spray dried *C. vulgaris* (56.7 ± 13.7%, 44.3 ± 7.5%, and 46.9 ± 12.7%, individually), in addition to fundamentally improving protein productivity proportion and nitrogen balance.

Also, there were no unfriendly impacts to the histology of significant organs upon long term utilization of ultrasound-treated microalgae, and along these lines, it has been recommended as a practical pre-treatment technique for the food business.

Altering two counter-current frequencies with each other has additionally been recommended as a suitable technique for additional improving protein extraction, as exhibited by a half increment in yield and 18% decrease in extraction time utilizing 15 and 20 kHz in *Porphyra yezoensis* contrasted with mono-recurrence UAE.

7.3.2. Pulsed Electric Field (PEF)

Pulsed electric field (PEF) has been utilized as a cell disturbance procedure in microalgae, in spite of the fact that its essential use has up to this point been for the extraction of lipids for change to biofuel. PEF includes applying high electric flows so as to puncture a cell wall or cell film, causing reversible or irreversible electroporation.

Electroporation empowers the introduction of different remote parts with cells, including DNA, proteins, and medications. PEF is a quick and green innovation for inactivating microorganisms by irreversible electroporation and supporting the arrival of intracellular contents of plant cells. In any case, conductivity, and anode gap are factors that might constrain this innovation for up-scaling.

Goettel et al. (2013) were among the first to report the utilization of PEF as a method for separating different intracellular segments from algae. From that point forward, PEF has been exhibited to build the yield of a few high-esteem microalgae parts, including lipids, sugars, carotenoids, and chlorophyll.

Protein yield from *Chlorella* sp. furthermore, *Spirulina* sp. was accounted for to increment by 27% and 13%, separately, following PEF-treatment at 15 kV/cm and 100 kJ/kg. Coustets et al. (2013) additionally announced essentially expanded protein extraction in *C. vulgaris* and *Nannochloropsis salina* following PEF-assisted extraction, taking into consideration the extraction of flawless cytosolic proteins (Figure 7.3).

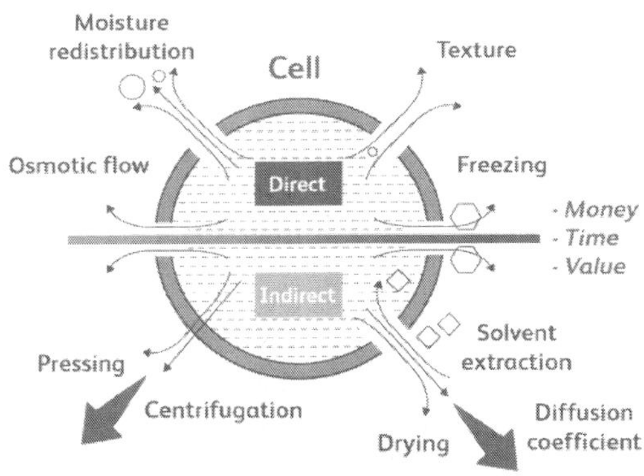

Figure 7.3. *Extraction by pulsed electric field.*

Source: Image by Pulsemaster.

7.3.3. Other Methods

Microwave-assisted extraction (MAE) includes warming a material, making moisture vanish, accordingly making bubbles under high pressure which would then be able to burst to cause disturbance in cell contents.

Expanded degrees of dissolvable proteins were separated from a microalgae biomass containing green microalgae (*Stigeoclonium* sp, *Monoraphidium* sp.) and diatoms (*Nitzschia* sp., *Navicula* sp.) utilizing microwave pre-treatment contrasted with ultrasound.

MAE has stood out for extraction of compounds due to low vitality proficiency, despite the fact that its utilization in algae might be constrained by debilitated capacity with dried samples (Figure 7.4).

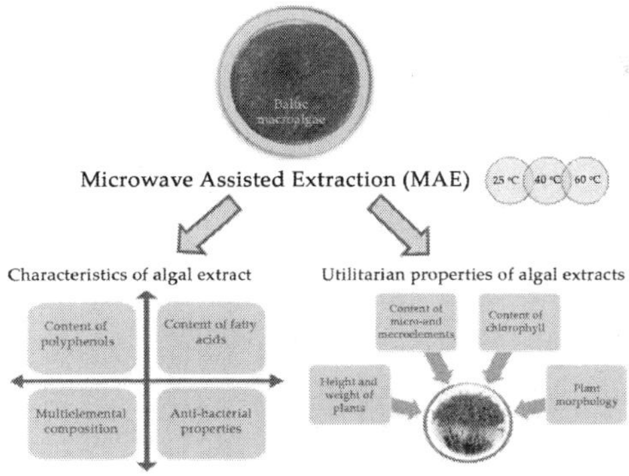

Figure 7.4. *Seaweed extracted by employing MAE.*

Source: Image from De Gruyter.

On the other hand, sub-and supercritical liquid extraction (SFE) procedures have picked up notoriety in late decades as extraction techniques. Subcritical water extraction (SWE) includes utilizing heated water (100–374°C) under high tension (~10 bar) to keep up water in a fluid state. On the other hand, SFE is a method that heats a liquid over its critical point, making it supercritical.

Under supercritical conditions, the properties of the liquid become undefined from its vaporous state, with a density like a fluid, however consistency coordinating that of gas. SFE ordinarily uses carbon dioxide (CO_2), making it a moderately 'green' innovation with low dissolvable utilization.

However, SWE, and SFE both require high venture costs for equipment and have regularly just been utilized in algae to date for the extraction of lipids.

7.4. ADVANCEMENT METHODS: MEMBRANE FILTRATION

Membrane technologies are broadly utilized in the dairy business to recuperate whey proteins from milk discharged because of the cheese making process. Membrane technology alludes to the utilization of a semi-porous membrane to isolate a fluid into two distinct parts by specifically permitting a few compounds to go through while blocking different compounds, regularly dependent on molecular weight.

Membrane technologies are promising elective strategies for enhancing algal proteins, in addition to creating novel techno-useful and bioactive ingredients. They have the upside of being non-thermal and good for the environment. The most generally utilized membrane technologies incorporate microfiltration (MF), ultrafiltration (UF), nanofiltration (NF), and reverse osmosis (RO).

Membrane advances could go about as an elective strategy for enhancing algal proteins when utilized related to a cell interruption procedure, for example, polysaccharides hydrolysis, UAE, or PEF. Interruption of the extreme cell wall is a basic step required so as to build the accessibility of algal proteins for extraction. Also, membrane technologies are appropriate for use with seaweed as a major aspect of a falling biorefinery procedure to boost valorization of all segments inside algae, while evading the presence of substantial metals in the last item.

A mix of membrane technologies could be utilized to disengage algal proteins utilizing similar standards of molecular weight cut-offs utilized in the dairy business. In the dairy business, MF is utilized to expand the timeframe of realistic usability of milk with no heat treatment by evacuating microorganisms, while protecting overall taste and tactile properties.

MF could be utilized to expel algae cell wall segments and microscopic organisms with a molecular weight more than 200 kDa. UF could then be utilized to segregate proteins and different macromolecules somewhere in the range of 1 and 200 kDa, like the manner in which it is utilized in the dairy business to create advanced parts under 10, 5, 3, and 1 kDa.

NF could then be utilized to expel monovalent salts to limit osmotic pressure, trailed by RO to diminish liquid volume. Undoubtedly, membrane technologies have just been utilized to separate entire microalgae cells and a few seaweed parts. Tangential flow MF was accounted for as a proficient technique for recouping 70%–89% of algal biomass from wastewater medicines.

UF was recently utilized related to supercritical CO_2 extraction and ultrasound to separate *Sargassum pallidum* polysaccharides. Polysaccharides with cell reinforcement exercises were confined from *U. fasciata* using hot water extraction followed by a few phases of UF with progressively smaller pore sizes.

Moreover, UF was utilized to segregate phycoerythrin protein from *Grateloupia turuturu* following cell homogenization, which was accounted for to hold 100% of the protein without denaturation.

On the other hand, a two-phase UF could be applied for algal protein advancement, as showed by the detachment of polysaccharide segments in *Tetraselmis suecica* utilizing diverse estimated pore membranes following high-pressure homogenization.

7.4.1. Environmental Risks Related with Creation of Seaweed

Development of kelp has been there all through Asia, and there is presently developing enthusiasm for the development of macroalgae in Europe to meet future resource needs.

On the off chance that this industry is to get built up all through Europe, at that point adjusting the related natural risks with potential advantages will be important to guarantee that the carrying limit of the affected environments are not surpassed and conservation objects are not subverted.

This is a precise audit of the environment changes liable to be related with a creating kelp aquaculture industry. Observing suggestions are made by chance positioning ecological changes, featuring the momentum information gaps and making way for research needs to address them.

Ecological changes of most prominent concern were distinguished to include: assistance of ailment, adjustment of population genetics and more extensive modifications to the local physiochemical condition.

Current elevated levels of vulnerability encompassing the genuine degree of some natural changes mean conservative risk rankings are given. Suggested checking options are talked about, that intend to address vulnerability and encourage informed decision-making.

While present small-scale development ventures are not considered that risky, an extension of the business that incorporates 'enormous scope' development will require a progressively complete comprehension of the scale dependent changes so as to offset ecological risks with the advantages that seaweed development endeavors can offer.

7.5. DRIVERS OF ENVIRONMENTAL CHANGE

7.5.1. Absorption of Light

It has been mainly demonstrated that the competition for light is important basically for the structuring of the aquatic algal communities. The above-mentioned understory was shaded by kelps (Reed and Foster, 1984; Clark et al., 2004; Flukes et al., 2014; Benes and Carpenter, 2015).

By the means of the water column itself, Morel (1978) and Platt et al. (1988) showed the light intensity and its quality as well as the by vegetation is indirectly altered by Reed and Foster (1984) and Clark et al. (2004). Further, it has been observed that when the terrestrial forests when compared to the giant kelp habitats is basically the vertical bottom up structure mainly (Dayton and Tegner, 1984).

As the kelp, canopies are basically capable of reducing the light that reaches the benthos less than 3% of the surface flux (Reed and Foster, 1984) in the same way the benthic shading by kelp can affect the understory algae.

It has been basically observed that the place where the light conditions are suitable for the growth there the natural macroalgae communities are mainly limited by such a habitat (Burrows, 2012).

As the crops basically must be cultivated in the surface waters at depths that optimizes levels of the photosynthetically active radiation (PAR), the aspect which has been observed was that the cultivated seaweeds habitats differ from the natural macroalgal.

The reduction in the photosynthetic efficiency could be mainly caused by the means of the excessive light which in turn causes the photo oxidative stress (Heinrich et al., 2012). The results such as the low levels of the photosynthesis and growth were mainly observed due to the levels of the PAR which basically are the suboptimal for the species being cultivated.

Further shade underlying habitats containing autotrophic organisms (e.g., pelagic phytoplankton and benthic macroalgae) was therefore occurred due to the cultivation of the seaweeds on the surface waters. Whether the cultivation systems overlap with the habitats containing the autotrophic species are the important aspects which are needed to be understood.

A modular approach which is similar to the other aquaculture activities explicitly requires the up scaling of cultivation practices in Europe. As in the specific areas the potential for breaking waves which causes the excessive wear and infrastructure failures are often observed and these places are rocky, shallow (e.g., less than 10 m) water environments are technically challenging places to deploy cultivation systems.

The longer mooring systems are possibly subjected to greater drag forces and they are also included with the Deeper (e.g., more than 60 m) water cultivation systems have other technical challenges. The shallow habitats supporting productive benthic macroalgae communities will be overlapped by the given of the above consideration which are unlikely that cultivation systems being developed in various developing countries.

When the possible sites were considered as such the species are afforded a high level of protection in various developed countries mainly in the case of Maerl beds and seagrass communities should be avoided and this is mainly sensitive to shading effects and or disturbance (Wilson et al., 2004).

These communities are therefore generally tolerant to periods of low light irradiances and also these communities are typically adapted to low irradiances. When sitting projects are being considered it has been mainly observed that the there is a possibility of negative benthic shading effects.

Mainly the significant detrimental effects are caused by the benthic shading at small medium and large scales and also the assuming that cultivation projects will have been limited overlapped due to the sensitive benthic environments.

Further, a high level of protection is mainly afforded by avoiding the habitats. As the cultivation, systems will be designed to efficiently absorb the irradiance at the water's surface mainly as the shading has the implications for the pelagic environment.

Further another aspect is that the where the floating kelp fronds reduce the irradiance in the upper meter of the canopy is mainly observed the similar shading in giant kelp community. Further due to the transmission through heterogeneous kelps blades (Gerard, 1984) a light penetration is exponentially related to the canopy density, but can be higher than they were expected to.

Further to limit macroalgal photosynthesis even under sunny conditions mainly in the well vegetated areas, an average irradiance at 1 m depth is low enough to limit this (Gerard, 1984).

The phytoplankton communities will only experience shading for the length of time it takes to travel through the site mainly apart from the possibility of shading effects on the sessile organism which are mainly located under the cultivation projects, and also water movement are required for the efficient nutrient and gas exchange among the cultivated species means this.

Further, it has been observed that at the small medium scales for individual cultivation, sites are highly unlike mainly for the significant shading effects on pelagic communities but on the other hand, it has been observed that it may act cumulatively within a site with multiple cultivation systems.

Further, an increase in the competition for the light from cultivated species is observed at large scales, due to the changes to plankton communities and the possible as phytoplankton will experience this increase.

The suppression of the abundance of the phytoplankton during the growing reason is mainly observed at the site of a large-scale kelp cultivation site in Sanggou Bay (Yellow Sea, China) (Shi et al., 2011). Further, by the means of affected marine food webs there are basically certain changes in the primary productivity which can in turn affect the trophic flow.

By using the Ecopath tom, model trophic structure was mainly use for the ecosystem structure which was studied in the areas of intense kelp cultivation (Wu et al., 2016).

Mainly by the provision of the habitat and food resources directly and indirectly through enhancing detrital biomass mainly the kelp cultivation in these areas was shown to have restricted trophic flow into the water column primary production and also for the benthic food (Wu et al., 2016).

A number of factors including the competition for nutrients, increased grazing pressure from epibenthic species, and altered hydrodynamics in addition to elevated completion for light are mainly the complex factors and also determining the causes and consequences of change in the phytoplankton community is also complex.

Under the marine strategy framework derivatives the phytoplankton communities have limited protection in a developed nation such as Europe also it is clear that there is an absolute requirement of the maintenance of the marine environment to good ecological status on the regional scale.

Further, the significant negative environmental effects at small-medium scales will be caused by the shading as it is sufficient enough to cause this change. Further, for determining the potential negative interaction at a larger scale therefore a focused monitoring program assessing the phytoplankton changes is required.

7.5.2. Absorption of Nutrients

Basically, the nutrients from a range of the natural marine and atmospheric sources are obtained by depending upon the connections with the ocean, coastal seas (Paerl, 1995; Prospero et al., 1996; Jickells, 1998; Baker, 2003).

Anthropogenic sources which are namely finfish aquaculture, agriculture, and urban wastewater) (Smith, 2003) are required for basically adding on to the nutrients to the marine environment by their means. Further, it has been mainly observed that the increase in the occurrence of the harmful algal blooms is the ones which are related to these sources of nutrient fluxes often (Anderson et al., 2008; Heisler et al., 2008).

Further, there are various negative impacts on the coastal ecosystems which have been altering the local ecology and the ecosystem services mainly due to the human-induced changes to the sources and the sinks of the nutrients (Shumway, 1990; Anderson et al., 2002; Heisler et al., 2008).

From the marine environment during the growth the seaweeds in suspended cultivation removes the inorganic nutrients (Kerrison et al., 2015; Marinho et al., 2015). The anthropogenic activities are mainly added along with the quantity and the proportion of nutrients which are being removed are equal to them and all of this effected by the means of the positive remedial effects (Seghetta et al., 2016b).

The proportionate amounts of the nutrients from the surrounding water body could be extracted by the means of the natural primary productivity and very large-scale culture of macroalgae and for this, the undesirable effects could occur if the nutrient removal by cultivation results in concentration (Lüning and Pang, 2003).

Further by increasing surface drag it has been observed that the suspended aquaculture systems used to cultivate seaweeds as it affects the local hydrodynamic movements.

Mainly reducing the water exchange necessary for maintaining levels of the nutrients which are mainly required for the growth (primarily Dissolved Inorganic Nitrogen) (Shi et al., 2011) also by the alterations to water flow can affect that carrying capacity of a water body.

In the current scenario, the models presented by Aldridge et al. (2012) are being observed in China where the nutrient depletion has been observed as the current realistic stocking densities in the developed countries as they are generally lower.

For example, in the most productive areas, China produces up to the 18 tons of dry kelp per hectare (Aldridge et al., 2012). Approximately 151 wet tons per hectare (10,000 m^2) (*Saccharina latissima* dry/fresh = 0.12) is mainly equated by mainly assuming the similar water content between kelp species.

Further it has been noted that throughout the growing seasons it should be noted that the dry vs. fresh weight (FW) vary (Broch and Slagstad, 2011; Peteiro and Freire, 2013a).) rope craft culture methods that are densely packed (stocking density estimated at approximately 0.66 linear meters of growing line per meter squared of cultivation area) (Shi et al., 2011) and the grid systems are employed in China either it is vertical or horizontal was preferred.

It has been observed that the reported growing systems in various countries are less efficient (Peteiro and Freire, 2013a). For example, if analyze the figure of China the above information infers that it is mainly able to produce 22.9 kg per line meter growing systems must first increase the density if seeded materials whilst increasing the yields per liner meter [average 9.1 kg wet weight per linear meter (Seghetta et al., 2016a)].

Proposing the potential for a huge decrease in local nitrogen assets on significant scale like extraction of nitrogen at 480 tons for each year for a site creating 20 t/ha dry load for a model of nitrogen necessities in a speculative huge farm (20 km^2) in the Clyde estuary in Scotland.

The nitrogen substance of dry material is roughly 1.2% at the time of an ordinary reap. Seeded material would be comparable to those seen in China (22.9 kg m^{-1} per line) even where we accept future profitability per meter of the seeded substance.

Extraction of 0.33 tons of nitrogen and production of up to 229 tons of biomass (27 tons dry weigh) would take place in small-medium scale operations (<50,200 m lines).

Expecting development rehearses are situated in zones with artificially elevated nitrogen assets during times when the harvest development rate is high at the previously mentioned scales negative environmental impacts from diminished nitrogen assets are profoundly impossible.

In the marine condition, anthropogenic wellsprings of nitrogen can be of a noteworthy sum. For instance, in 2010 a measure of 7500 tons of nitrogen is being assessed to be discharged by the Scottish salmon farming industry in the marine environment (Aldridge et al., 2012).

Negative environmental impact of limited nitrogen consumption might be kept away from if cautious thought is given to the siting of ocean growth ranches, in this way guaranteeing conveying limit of the conditions are not surpassed as the earth overshooting day getting prior every year.

On the off chance that co-situated in appropriate zones of high anthropogenic nitrogen input remediation of overabundance nitrogen at bigger territorial scales like development tasks, cultivation projects may contribute generously.

Seaweed cultivation development at a large scale has a constructive outcome through bio extraction of N and P from anthropogenic exercises in the marine condition and help management techniques at the water body level indicated by a life cycle impact assessment (LCIA) of seaweed cultivation and nutrient extraction in Europe (Seghetta et al., 2016b).

For the advancement of nitrogen mass balances for every cultivation or development site, the planning of profluent discharge and take-up, dispersal attributes of the site alongside, and information on nitrogen cycling in the environment is required.

The executives of supplement levels and permitting of the scaling cultivation undertakings to the attributes of the particular water body would be bolstered by such information in an increasingly comprehensive methodology.

Negative interactions associated with a development framework, cultivation infrastructure may reduce the general advantages of such a methodology if ineffectively sited and present seaweed biomass required to expel the nitrogen gushing from an ordinary salmon farm is a lot more noteworthy than small-medium seaweed activities can deliver (approximately 1000 wet tons) (Broch and Slagstad, 2011).

Therefore, by giving greater clearness to evaluated sources and sinks of nitrogen the advancement of coupled hydrodynamical-biological models at industry sensible stocking densities will bolster future turns of events.

In areas where algae growth development is a rapid phenomenon and regular restoration of nitrogen assets is influenced by changed water trade subsequently, the competition between developed algae growth and phytoplankton can be normal at time stretches in the creation pattern of the equivalent above region. Exhaustion of the phytoplankton communities could have negative ramifications for certain species in influenced regions where activities are the enormous scope and have high stocking densities.

A level of site-explicit displaying and checking work is required to guarantee a solid proof base to decide the exchange offs and cooperation's related with enormous scope macro-algae creation as opposed to securing, moderating, and improving biodiversity for the achievability of the huge scope cultivation venture.

7.6. CHALLENGES IN EXTRACTION PROCESS

7.6.1. Access Rights

At present, all seaweed species are gathered manually in Ireland. This incorporates the most financially important species brown seaweed *Ascophyllum nodosum* and the red calcified coralline seaweed *Phymatolithon purpureum* and *Lithothamnion corallioids* (usually called maërl).

The enactment for collecting seaweed in Ireland depends on the Foreshore Acts 1933–1998, which expresses that the Department of Communications Marine and Natural Resources is enabled to give licenses for seaweed reaping on the seabed out to 12 nautical miles.

There are right now no limitations on collecting amounts in Ireland, nor are there any limitations on reaping times. The main exception to this is for maërl because of its amazingly moderate development rate (0.6–1.5 mm per annum).

In any case, the future usage of mechanical harvesting strategies will probably require a survey of existing enactment to guarantee that fitting access rights and feasible collecting yields are kept up. Mechanical harvesting of *Laminaria* spp. is now present in different European nations, for example, France, and Norway, giving a significant resource to making any such revision.

7.6.2. Variability

An extra challenge, which is especially pertinent in the creation of bioactive peptides, is the significant levels of fluctuation in algal proteins. The protein substance can change via season, temperature, and area in which the seaweed is collected.

The general composition of specific proteins inside the plant can likewise contrast, changing the concentrations of amino acids and thus modifying the yield of wanted peptides as a result. For instance, yearly monitoring of *P.*

palmata harvested on the French Atlantic coast demonstrated that protein levels were most elevated in the winter and spring months, changing from 9 to 25%, and reaching its peak in May.

Thus, *Gracilaria cervicornis* and *S. vulgare* differed via season, with protein levels contrarily corresponding with temperature and saltiness. Diverse harvest areas of *U. pinnatifida* in New Zealand likewise altogether influenced the protein substance and amino acid creation.

7.6.3. Scalability

The scalability of protein extraction from algae is a further hindrance that should be defeated before seaweed and microalgae become a feasible source. Algal protein extraction is still especially in its earliest stages, implying that a significant number of the strategies that have been created are still at small scale.

PEF and ultrasound have been recommended as being appropriate for enormous scope algal protein extraction. Membrane technologies may likewise be versatile for business applications, with UF detailed as being appropriate for R-phycoerythrin extraction from the seaweed *G. turuturu* at the industrial scale.

7.6.4. Digestibility

One of the most significant difficulties for removing proteins from algae is the elevated levels of cell wall anionic polysaccharides which can get bound to the proteins and make viscous medium, further increasing the challenge of protein extraction.

The morphology of various seaweed species has been proposed to be a significant factor in deciding protein yield, with harder thallus structures believed to require more processing.

A few brown algae species (*Eisena bicyclis, H. fusiformis, U. pinnatifida*) were accounted for to have diminished *in vitro* digestibility (51.8–57.1%) contrasted with red algae P. palmata and *P. tenera* (70.2% and 87.3%, separately), which was recommended to have been because of expanded degrees of dietary fiber.

Moreover, the kind of polysaccharide has been appeared to impact the degree of protein absorption. *P. palmata* and *G. verrucosa*, both red seaweeds, were appeared to have fundamentally changed degrees of *in vitro* revised nitrogen digestibility (1.9% and 16.3% after 6 h, individually).

This was to a great extent ascribed to contrasts in polysaccharide divisions (pentoses and hexoses, respectively). Medicines to disrupt the cellulosic cell wall could assist with defeating this issue to make algal proteins and other cell segments increasingly available.

Warmth treatment is one such alternative that can improve food taste, surface, wellbeing regarding allergenicity and microbial burden, and conservation, while additionally increasing bioavailability and usage of proteins by incomplete denaturation and breakdown of proteins into peptides, taking into account simpler access by proteolytic enzymes. A case of this can be seen with bubbling of *P. palmata*, bringing about a 64–96% expansion in freed amino acids.

Protein digestibility can likewise be to a great extent affected by the presence of phenolic compounds present in algae, particularly phlorotannins in brown seaweed. Oxidized phenolic compounds can respond with amino acids to shape insoluble complexes, which may hinder proteolytic proteins and therefore decline their nutritive qualities.

7.6.5. Food Safety

While algae can be characteristic aggregators of nutrients and minerals, they can likewise gather harmful components, for example, heavy metals. There are accordingly severe lawful cutoff points in Europe for the protected maximum presentation of heavy metals, for example, mercury, arsenic, lead, and cadmium, in foods for human utilization.

These cutoff points depend on the proposals of the Joint FAO/WHO Expert Committee on Food Additives (JECFA) and the Panel on Contaminants in the Food Chain of the European Food Safety Authority (CONTAM Panel), and are sketched out in the enactment under Commission Regulation (EC) No. 1881/2006.

The absence of guideline for algae as a dietary enhancement is of extra concern, prompting huge varieties between various clusters from a same producer. This has prompted concerns being raised about quality and defilement at the production and processing stages.

Downstream applications for which the algal proteins are expected can assume a job in deciding processing techniques.

For instance, food grade reagents and procedures might be required for proteins expected for human utilization, though this may not be such an issue with proteins planned for business applications or creature feed.

Moreover, any reagents utilized for the creation or extraction of algal proteins should likewise be environment friendly with least effect on local biological systems.

7.6.6. Price

A basic factor that will decide the business reasonability of microalgae and seaweed is their competitiveness contrasted with different sources available. For instance, a significant application for the creation of microalgae is the extraction of lipids to be changed over to biofuel.

However, while they are a greener option for nature, they are as yet not competitive contrasted with fossil-based petroleum products. Algae power could be additionally increased by taking an all-inclusive perspective and boosting the extraction of all accessible high-esteem parts by falling biorefinery.

Customary strategies for algal protein extraction are frequently very inefficient as they normally discard the algae results following processing. Seaweed is obviously appropriate for falling biorefinery, on the grounds that it contains some high-value components in addition to massive low-esteem segments that are viewed as crude materials for the bio-based economy, for example, xylose, and glucose.

As a few studies have appeared, the results of algae can in any case have numerous applications that could be of financial worth.

7.7. CONCLUSION

The extraction of seaweeds is a typical process and involves a lot of methods and effort. The processes involve both conventional and modern approaches and have been discussed in detail. The processes highlight the role of light and the proper extraction of the nutrients in the process.

There are various challenges that may arise in the process of extraction such as those of the rights to access the seaweeds, the digestibility of the seaweeds, the extent of food safety, the price that must be set on the products and the cost incurred on the extraction, the scalability, and the variability.

REFERENCES

1. Bleakley, S., & Hayes, M., (2017). Algal proteins: Extraction, application, and challenges concerning production. *Foods*, [online] *6*(5), p. 33. Available at: https://pdfs.semanticscholar.org/23bf/8fd65fc4c9b10fa47cd4ddb45cd13b1aa01c.pdf?_ga=2.864369.1150844088.1592460988-1723960050.1592460988 (accessed on 10 July 2020).
2. EL Boukhari, M., Barakate, M., Bouhia, Y., & Lyamlouli, K., (2020). Trends in seaweed extract-based biostimulants: Manufacturing process and beneficial effect on soil-plant systems. *Plants*, [online] *9*(3), p. 359. Available at: https://www.mdpi.com/2223-7747/9/3/359/htm (accessed on 10 July 2020).
3. Fao.org. (n.d.). *Production, Trade and Utilization of Seaweeds and Seaweed Products*. [online] Available at: http://www.fao.org/3/ac860e/AC860E01.htm (accessed on 10 July 2020).
4. Ganesan, M., Trivedi, N., Gupta, V., Madhav, S., Radhakrishna, R. C., & Levine, I., (2019). Seaweed resources in India-current status of diversity and cultivation: Prospects and challenges. *Botanica Marina*, [online] *62*(5), pp. 463–482. Available at: https://www.degruyter.com/view/journals/botm/62/5/article-p463.xml (accessed on 10 July 2020).
5. Kim, J., Stekoll, M., & Yarish, C., (2019). Opportunities, challenges, and future directions of open-water seaweed aquaculture in the United States. Phycologia, [online] 58(5), pp. 446–461. Available at: https://www.tandfonline.com/doi/full/10.1080/00318884.2019.1625611 (accessed on 10 July 2020).

CHAPTER 8

Market Trends of Seaweed Products

CONTENTS

In this chapter, there is a discussion about some general trends about seaweed. Some other important market trends are also mentioned.

In this chapter, the market dynamics have also been discussed. The detailed market and trade trends about various geographical regions are discussed in this chapter.

In this chapter, the market outlook and the need for global halal certification.

8.1. INTRODUCTION

Many health advantages associated with a natural source would raise the market share of the industrial seaweed market over the predicted timeframe. In pharmacy, food, and numerous other applications, numerous forms of eatable seaweed can be utilized. The commodity has the strongest iodine and calcium supply that prevents the vitamin A and C shortage.

The sustainable options for industries got popular due to growing climate change concerns and growing awareness of requirement to conserve resources. The rising need for eco-friendly fertilizers is the answer to the growth of the size of industrial seaweed business sector. Due to result of massive use in manufacturing of fertilizers development in the coming future, it is expected to impact demand of industrial use seaweed.

The demand for organic pulses and other food products would increase a major shift in customer preferences to healthier consumption, which would greatly influence product demand. The high content of organic matter, micronutrients, enzymes, minerals, and fatty acids found in fertilizers produced from seaweed are more beneficial compared with the chemical fertilizer. This will contribute significantly to improving agricultural production and boost the market size of commercial algae.

In tackling major problems and bettering the health of consumers, the nutrition provided by seaweed is very beneficial. It is ideal for use in the medicinal sector, the pharmaceutical industry and it is in huge demand by the manufacturers and suppliers, thanks to its antifungal, antiviral, and antibacterial properties.

The increased volume of biofuels produced by seaweed is one of the world's tremendously growing business sectors. Many research and development activities to establish effective biofuel-generating technologies will boost demand for seaweeds and affect market outlooks for industrial level algae.

It is expected that red seaweed is expected to have one of the fastest growth levels during the upcoming years. This item is widely used in Japan, China, and Korea as additives for soups. Its use as an additive will drive its demand in the food industry. In the medical and the pharmaceutical sector, Brown seaweed has shown considerable demand. The increase in the use of this type of products type will surely increase the market share of the seaweed industries and in many applications, such as food additives, pharmaceuticals, and products.

Thanks to increased acceptances of the market in relation to liquid type in numerous vertical industry components, like the textile, food additives, pharmaceutical, and medical products, direct food, and others, this product sector CAGR is approximated to grow by 8% over the next few years.

The most important application of liquid seaweed is to generate fertilizers that render them more nutritious and competitive, contributing to the production of agricultural goods. A higher demand for fertilizers merged with seaweed has required increasing agricultural activity worldwide, which in turn will trend toward the liquid market of commercial maritime algae.

In some Asian countries, Asia Pacific is a major contributing factor to the world algae market due to the massive demand for the product. For rice balls, pancakes, sweets, and preparing soup countries like China, Japan, and Korea like using seaweed. The product has significant health benefits and medicinal products which can treat wounds. It is predicted that APAC's commercial seaweed market is growing in a robust 8% CAGR from 2017–2024.

Some major market players involved in the production and distribution of commercial seaweeds are companies such as Chase Organics, Acadian Seaplants Ltd., CP Kelco, Pacific Harvest, Indigrow Ltd., Mara Seaweed and Aquatic Chemicals.

In 2019, the world seaweed industry achieved 8.8 billion US dollars in value of manufacturing. The different varieties of macroscopic, multicellular, and marine algae that live and grown the oceans, lakes, rivers, and other bodies of water are known as seaweeds or macroalgae. They are collected either by aquaculture or by wild harvest.

The different hydrocolloids, like the agar, alginate, gelatinous substances, and carrageenan can be derived from them as per the high level of minerals and vitamins contained in seaweed. It is used in many different products and sector like in food and beverage products, pharmaceuticals, wastewater management, bio-refining, and dietary products because of variety of benefits related to health and body function.

The requirement for algae is increased by the use of seaweed in dietary supplements on the account of their nutritious value, flavors increase, and weight loss management. In addition, seaweed is rich in bioactive chemicals, antioxidants, and antimicrobial properties, making them ideal in pharmaceutical industries for many medicinal purposes.

In the past several times the demand for products has also been supported by new uses of seaweed, for example as a thickener in the textile industry as well as for improving yield, enhancing land qualitie,s and protecting crops in the farming industry. In addition, the increasing investment in research and development activities in conjunction with the increased awareness of consumers on diverse marine algae applications is expected to catalyze the world algae market in the coming years (Figure 8.1).

Figure 8.1. Seaweed is an important source of food in many Asian countries.

Sources: Image by Pixabay.

8.2. SOME IMPORTANT TRENDS

8.2.1. Green Seaweeds Is Projected to Be the Fastest-Growing Segment during the Forecast Period

The use of green seaweed is in both human food and other industries are widespread. In countries like the Philippines, green seaweed like the caulerpa and lens is eaten as fruit. In the projected time period, they would be used for other fields of operation such as biofuel generation and wastewater treatment initiatives from government are expected to stimulate market growth.

> For example, UK Transport Department initiatives are expected to have a positive effect on green seaweed demand to increase biofuel consumption by 2020 (FAO).

For example, UK Transport Department initiatives are expected to have a positive effect on green seaweed demand to increase biofuel consumption by 2020. The cleaning of toxic substances from industrial wastewater is done with green seaweed like the Ulva spp. Green algae are also used as an instrument for bioremediation.

8.2.2. Food Is Projected to Account for the Largest Market Share During the Forecast Period

From ancient times, in many countries' seaweed has been used in countries as a food additive like in Japan, Korea, and China. Because of the availability of nutrients like high iodine, calcium, magnesium, iron, vitamin, antioxidant, and fiber, it is commonly acknowledged to be healthy food. In large continents like North America and Europe, the seaweed market has been guided by this common belief as a food source. The mainly consumed algae in the food sector are Nori, Kombu, and Wakame.

8.2.3. North America Is Projected to Be the Fastest-Growing Region in the Commercial Seaweeds Market, in Terms of Value

Due to the rising usage of commercial seaweed in the food industry, North America is projected to be the fastest-growing market in the coming years (FAO). The area, though, is controlled mainly by the United States. The market for commercial marine algae in North America is motivated by several factors including growth in the agricultural usage of algae and expanded understanding of the functional properties of marine algae to improve nutritional content in foods and their use as food supplements.

8.3. IMPORTANT SEAWEED MARKET TRENDS

8.3.1. Higher Benefits with Enhanced Yield and Productivity

Agriculture is vital to people's wellbeing and the development of the world economy as a whole. Dietary requirements and production criteria are increasingly relevant in the future because of the rising global population. This can be done by leveraging a number of emerging inventions and techniques to promote agriculture; a few of includes the use of seaweed extract in order to improve agricultural growth.

The derivatives from liquid seaweed are versatile and have shown different pathways to increase efficiency. Additionally, changes to root systems, enhanced plant growth, such as flora and fruit production, and an enhanced capacity to withstand plant disease and temperature stress, such as cold and dryness, are also advantageous. These also strengthen the soil composition, the retention of soil water, and the microbiological growth of soil systems.

8.3.2. Asia-Pacific Held the Largest Share of the Commercial Seaweed Market

In Asian region countries like China, Japan, and Korea usage of seaweed as food has deep origins, as well as in North America, South America, and Europe have witnessed the sharp increase in demand for seaweed. The Republic of Korea and Japan are second and third and china is the top manufacturer of seaweed till now. Seaweed though is now produced in all the major geographical regions.

There is a strong growth trend of seaweed in the United States, Mexico, and Canada. The demand for Carrageenan is anticipated to rise by 50% per capita in the coming years because of the huge demand by consumers. The largest seaweed products market is the Asia-pacific region. It's the largest global market in commercial seaweed production and use.

8.3.3. Competitive Landscape

The demand for commercial seaweeds is widely decentralized in nature and competitive in the markets around the world. The most prominent companies in all key players on this market are Cargill Inc., DuPont, Acadian Seaplants Limited, CP Kelco, Algae, the Arctic group, Fruit Hill Farm and Kelpak. The companies like Cargill Inc. and CP Kelco are the major firms in this sector that are in developed countries.

In the developing Asia-Pacific nations, the European zone and North America, the bulk of main players are engaged in this sector. In many countries of South America and eastern Europe, the Middle East and the

Oceana region, their subsidiary companies and divisions are also formed and working at the same time (Figure 8.2).

Figure 8.2. Farmer harvesting the seaweed.

Sources: Image by Wikipedia.

8.4. MARKET DYNAMICS

8.4.1. Increase in Seaweed Harvesting to Cater to the Growing Demand for Seaweeds and Seaweed-Based Products

The extraction of marine algae was a longtime activity in Europe, and it has two main uses, to be used as food by people and as manure in agriculture. The growth of the industrial sector using seaweed has dramatically altered the working of the European coastal population.

In many European countries such as Norway, France, and Ireland have seen strong growth in the seaweed industry that time. The two major harvesting methods were adopted by European farmers, mechanical harvesting, and hand harvesting respectively. Almost all of the seaweeds are being used in farming and horticulture within these countries.

Thanks to their rich protein content and nutritional properties, many varieties of seaweeds are used for human consumption. These utilizations of seaweed have led to an increase in demand amongst European consumers for eatable seaweed.

The region of North America is already seeing a growth in the harvest and use of seaweed for the production of industrial products, in addition to the increasing requirement for seaweeds in countries of Europe. The seaweeds are primarily used for food and manufacturing processes like the paper pulp, nutraceutical, medicine, cosmetics, personal care and other goods in Northern America.

The other type of main uses in this area is the use of seaweed to produce hydrocolloids, like the alginates, agar, and carrageenan. Furthermore, because of so many benefits in the organic fuel industry, like the fuel security, carbon savings, and rural economic growth, the use of marine algae is also receiving increasing attention. The growth of the commercial seaweed industry has been further increased.

In addition, various research and development activities are undertaken in the region to discover different uses of seaweed, like the bioplastics, and preparation of food products which contain low fat, minimal calories and less saturated fatty acids which are expected to fuel commercial seaweed consumption around the world.

For example, seaweed is capable of supplying different types of fuel for the generation of heating and electricity, the transportation fuel, biodiesel, and biogas. In Europe, the United Kingdom is setting the target of using 10% of electricity generated by renewables in road and railways by 2020. This will theoretically affect the development of seaweed in the segment of biofuels.

8.4.2. Excessive Use of Seaweed Products Can Have Harmful Effects Due to the Presence of Toxic Minerals

The use of products of seaweeds and aquatic dried goods obtained from aquaculture has some adverse consequences, as radioactive minerals are present and causing health issues around the world. The commercially available seaweed-based products are partly dried aquatic algae packed in plastic containers and distributed in the form of tablets and powder.

The concentration of excess metal contaminants, like the arsenic and iodine, has raised health issues for seaweed collection from wildlife. The daily ingestion of such heavy metals in seaweeds may contribute to medical risks such as cancer, neurological injury, gastrointestinal disorders, and kidney failure.

Therefore, excess algae intake raises the iodine amount above the necessary level. This increases thyroid hormone levels and thus leads to severe conditions like the goiter disease. Such safety issues induced by the ingestion of algae products have limited seaweed ingestion and thus hampered this industry growth.

8.4.3. Increase in Technological Development to Cultivate Seaweed Would Reduce Labor Engagement

Across Asia-Pacific, the production of seaweed is mainly carried out. The seaweed is cultivated and shipped for further research to developing regions such as Europe and North America to produce extracts for end-user industries. As a consequence, owing to technical limitations in Asia Pacific, the exposure to labor is higher, that increases the operating costs. Thus, the introduction of emerging technology may be applied in different Asian countries to minimize labor expenditure, like the production of single plants in tanks on land.

8.4.4. Lack in Financial Support and Government Engagement and Improper Marine Spatial Plans Are Major Market Challenge to Seaweeds Market

The local small level seaweed growers have also faced major difficulties with advances in seaweed production. In the seaweed industry, broad market channels and income distribution are essential factors. In addition, the lack of financial support for local small level seaweed farmers by banks and financial institutions has formed operational problems.

In order to fulfill their financial and operating needs, they have to depend upon several other financial communities. This is an impediment to several future producers of seaweed, in particular small and medium-sized farmers of seaweed. Therefore, to resolve this problem, state involvement is crucial for the financial and infrastructural support for seaweed growers in a way to improve the output of seaweed.

With regard to marine spatial planning, most tropics industries are based on coastal areas and space disputes are numerous. The floating rack networks in the Northern Hemisphere should be introduced in tropical coastal regions in order to meet the growing need for seaweed.

The floating rack systems have been tested in strict field conditions with the capacity to yield 12 tons of dry seaweed per unit as per data from the World Bank group (Figure 8.3).

Figure 8.3. Red seaweed is in much demand in the international market.

Sources: Image by Wikimedia.

8.5. SEAWEED MARKET OVERVIEW

The seaweed industry around the world in 2024 is expected to hit $ 9,075.65 million by 2024, with a CAGR of 12.1% from 2018 to 2024. This amount was valued at $4,097.93 billion in 2017.

The seaweed was cultivated only by the coastal communities well before emergence of large-scale production of seaweed. Increased understanding of the safety advantages of seaweed and their industrial usage in specific end-use sectors has contributed to the introduction of industry activities in numerous countries around the world.

The growth of demand by consumers for marine algae as human food and in production of hydrocolloid (for example, agar, alginate, and carrageen) is predicted to increase demand for seaweed during the coming years. Seaweeds are used in fertilizers, animal feed, biofuels, and wastewater, that gives more support and complement the growth of the seaweed sector.

In certain countries in North America and Europe, as opposed to Asia-Pacific, the marine farming industry has grown while it's already in its infancy. The Blue Growth initiative of the European Union (EU) has led to the establishment of seaweed cultivation in the last some years. The countries in Europe are supporting this initiative by providing subsidies to examine the possible applications of seaweed in different types of industrial sector.

The expanded usage of red seaweed is projected to fuel development of the seaweed sector in the food industry. Moreover, in 2017, the human food sector had the largest market share of algae, since seaweed are consumed as important source of human food in many coastal countries around the world. In the world seaweed market,

Asia-Pacific held the world's largest market share and is expected to remain predominant in the predicted period. A significant percentage of the population from Asia-Pacific resides through coastal areas and is thus strongly linked to the sea and its food supplies. As a result, seaweed farming was established in Asia-Pacific. The country which is largest producer of seaweed is china and has a share of roughly one-fourth by value in the international market.

The pigments in algae are red, with phycoerythrin and phycocyanine masking other pigments like chlorophyll, beta-carotene, and several uniquely xanthophyllic. In the 2017 the red seaweed, which accounted for a share of more than eight-ninths in context of value and has dominated the international seaweed market in 2017.

The green seaweed produces herbal beta carotene is effective in avoiding other tumors, particularly the lung cancer. In fact, in the food industry green seaweed has many uses. The claim that red algae are the highest and richest of seaweed, is due to this. In addition, the availability of phycocolloids such as carrageenans makes them economically crucial in Asia-Pacific.

The seaweed produces a carrageenan that is used as a thickening agent and stabilizer in a number of food processing applications. In the food sector, thus, increased intake of red seaweed is expected to lead the growth in market.

In many coastal countries human beings have consumed and treated seaweeds as a source of food, hydrocolloids, fertilizer, animal food additives and other sources around the world that also comprises aquaculture, fuel biomass and wastewater treatment.

The human beings used seaweeds as algae are used as human food supply in several coastal countries around the world, the human food product category obtained the largest market share in 2017. However, in Asia-Pacific, in specific China, Indonesia, the Philippines, Korea, and Japan, seaweed farming has been a huge industry, is the main use of such seaweeds as food ingredients and in food.

The application of seaweed as food and animal feed in the majority of western countries has been prohibited and therefore there is no considerable pressure on developing seaweed growing technology. But as Asian-Pacific people moved to other regions of the planet, demand for algae as a food in different parts of the world has grown.

Over the projected period 2020 to 2025, the commercial algae demand in international market is predicted to reach 9.7% CAGR:

- Because of its workable and healthy benefits, there is a broad application base of commercial seaweed. The bulk of the commercially grown seaweed is used in the food sector, responsible for 70% of the total market share and 30% in the other segment, which contains hydrocolloids, supplemented by feed and fertilization.
- Seaweed markets exploded thanks to demands for weight reduction. In stomach seaweeds moisten also, swell up. Overall, algae are now becoming common not only as part of their staple diet, food flavor enhancers and nutrient products, but also as part of their properties to lose weight. The seaweed intake in the United Kingdom alone has increased by 125% due to this reason.

8.6. MARKETING AND TRADE

The approximated exported value of USD 4 trillion for 2016 was 1 million tons (product weight) of seaweed that is dried and processed. The 100 countries market around the world imports it and the key trade centers are Asia, Europe, and North America.

The publicly available data assessed by experts showed that in 2016, the top 35 countries accounted for almost 80% of global seaweed trade. In combination, the declared value at custom is nearly 650,000 tonnes (Product Weight) of dried seaweed, carrageenan, and agar that year was USD 2 billion. This volume also accounts for 70% of farmed seaweed production around the world.

So, it comes as no shock that marine aquaculture output has nearly doubled from 14 million tons in 2006 to nearly 30 million tons in 2015, attributed to its growing usage and business around the world. The top four import markets were China, the EU, Japan, and the United States of America (FAO) (Table 8.1).

Table 8.1. Seaweed Imports into the Top 35 Countries, 2013–2016

	Weight in Tonnes; Value in USD Thousands					
	2014		**2015**		**2016**	
	Weight	**Value**	**Weight**	**Value**	**Weight**	**Value**
Edible seaweed	247,527	692,978	251,709	633,915	250,735	648,719
Non-edible seaweed	244,514	348,059	244,777	271,209	245,381	269,473
Carra-geenan	115,467	913,486	111,852	891,776	110,555	759,844
Agar	11,738	243,316	11,771	231,958	12,052	225,624
Total	**619,246**	**2,197,839**	**620,109**	**2,028,858**	**618,723**	**1,903,660**

Sources: FAO.

The hydrocolloid products are most dynamically imported, and in many countries for different applications are widely traded and used. The majority of material suppliers are from d developing countries of seaweed for further processing, especially for the production of agar, alginate, and carrageenan. However, commerce is carried out almost entirely in Far East nations, including China, Japan, the Republic of Korea, and the island of Taiwan, for seaweed edible for human consumption.

In 2016, the dry seaweed has 50% share in the total seaweed imported into the international market, with equal share of edible and non-edible forms in commerce. The demand for imports of eatable sewage for direct consumption (e.g., wakame, nori) from eastern markets in Asia, including Japan, the Republic of Korea, and the Province of China of Taiwan was reliable. However, international trade in non-edible seaweed is displaying a downward pattern due to its increased use in the producing countries themselves in semi-processed and processed forms.

In fact, China has become the main importer for more processing of edible and non-edible algae, impacting global selling prices for dried algae. Furthermore, the EU and the United States of America remain the major destination for dried algae, in specifically for inedible raw materials for further production of carrageenan and for other agricultural good.

The dried algae raw material to European conventional markets have started to decrease, because China has become a major center for reprocessing, and many developing countries like the Philippines and Indonesia have started to develop their own agar and carrageen processing industries (Table 8.2).

Table 8.2. Top Four Importers of Dried Seaweed and Hydrocolloid Products, 2014–2016

	Weight in Tonnes; Value in USD					
	2014		**2015**		**2016**	
	Weight	**Value**	**Weight**	**Value**	**Weight**	**Value**
China	257,959	377,389	234,424	250,727	252,683	255,775
European Union	154,924	725,090	167,170	688,326	178,467	613,162
Japan	51,880	296,750	48,705	272,589	49,718	312,749
United States of America	38,102	256,442	38,278	253,227	40,861	209,143
Total	**502,865**	**1,655,671**	**488,577**	**1,464,869**	**521,729**	**1,390,829**

Source: FAO.

8.6.1. China

In the international seaweed trade, China is now the largest and most important partner, especially in the capacity of volume. China also has developed as the top re-processing agent for hydrocolloid products made of edible and inedible seaweed over the past decade. In the year of 2016 imports were 250,000 metric tons of dry raw material, in comparison to 2012 it is 20% higher.

The countries like Indonesia, Chile, and Peru were the major producer of the non-edible type; Indonesia is the top supplier of almost 90% of eatable, dried seaweed imports. China has now become a main stakeholder for deciding price for the trade in dry seaweed due to its dominance in the industry.

8.6.2. European Union (EU)

In 2016 with almost 180,000 tons of seaweed products were purchased, and it's an estimated 613 million US dollars in value, in context of volume the second largest market that import seaweed products is EU. The EU is therefore the top demand sector for manufactured seaweed in the world.

In 2016, non-edible dry seaweed rated number one in view of their quantitative imports, with expanded supplies to the EU. Denmark, Germany, Portugal, and Spain were the leading importers. The carrageenans were also the most important import value at 400 million US dollars in 2016 like previous trend witnessed in it. The countries like Spain, Denmark, France, Germany, and the United Kingdom were the major importing countries for these products.

In all, in 2016, carrageenan, dry seaweed, and agar valued a little less than in 2012 were imported by the EU of 613 million US dollars. But during the 2012–2016 periods, the quantity increased by 31% due to further seaweed imports. The top importers in the EU were France, Germany, Poland, and Spain.

In the case of carrageenan and high-quality agar agars, the EU also constitutes as a significant re-processing and export region. About 100,000 tonnes, the majority of which contained good quality processed algae, shipments in- and outer-EU seaweed that is processed and is of high quality (Table 8.3).

Table 8.3. European Union: Imports of Seaweed, 2014–2016

	Weight in Tonnes; Value in USD Thousands					
	2014		**2015**		**2016**	
	Weight	**Value**	**Weight**	**Value**	**Weight**	**Value**
Edible seaweed	22,953	75,813	17,175	60,726	15,184	59,662
Non-edible seaweed	65,310	74,957	89,103	79,962	88,485	69,496
Carrageenan	63,302	493,021	57,158	474,686	70,627	400,249
Agar	3,359	81,299	3,734	72,952	4,171	83,755
Total	**154,924**	**725,090**	**167,170**	**688,326**	**178,467**	**613,162**

Sources: Eurostat.

8.6.3. Japan

In Japanese food and cooking seaweed are customarily an essential element and thus the majority of imports into the country are for consumption by humans, one way or another. The biggest importer of edible seaweed likes Japanese wakame or nori, for that the average import value is USD 6–7/kg and in the scenario of reprocessing products in Asia or Europe is USD 1 per kg.

In general, imports account for approximately 35% to 40% of overall seaweed market. All seaweeds are packed and sold in dried form such as Laminaria spp., Porphyra spp., and Hizikia spp. (Japanese Kombu, Nori, and Hiziki, respectively). Undaria spp. is packed and sold in the boiled, salted form (Japanese wakame).

The yearly imports of marine algae into Japan for the duration from 2012 to 2016 ranged from 50,000 tons to 55,000, estimated at 300 million US dollars to USD 600 million dollars.

The important suppliers of edible algae were Chile, China, and the Republic of Korea.

The Southeast Asian countries such as Indonesia, the Philippines and Thailand supply carrageenan, along with Denmark and the United States of America, and the top suppliers of agar are Chile, China, Korea, and Morocco. The two latter sources have produced high-quality products (Table 8.4).

Table 8.4. Japan: Imports of Seaweed, 2013–2016

	Weight in Tonnes; Value in USD Thousands							
	2013		**2014**		**2015**		**2016**	
	Weight	**Value**	**Weight**	**Value**	**Weight**	**Value**	**Weight**	**Value**
Edible seaweed	30,776	177,688	31,427	179,062	30,182	164,276	32,989	210,207
Non-edible seaweed	14,548	36,058	15,141	36,630	12,903	28,089	11,497	27,605
Carrageenan	3,233	27,505	3,462	31,185	3,665	30,421	3,419	29,490
Agar	1,785	46,519	1,850	49,873	1,955	49,803	1,813	45,447
Total	**50,342**	**287,770**	**51,880**	**296,750**	**48,705**	**272,589**	**49,718**	**312,749**

Sources: Ministry of Finance, Japan.

8.6.4. United States of America

In semi processed and processed seaweed, the major producer is the United States of America and it continues to remain at top. Even though, there the reprocessing industry that needs to import non-edible seaweed for production of products of non-food quality like alginic acid.

In the United States of America imports of algae, carrageenan is the key ingredient. In the late decade of 1980, as the ban is removed on the Philippines Natural Grade (PNG) carrageenan from the Philippines by the United States of America, in the next five years the imports of products almost doubled to about 10,000 tons a year. It has stayed in some years at this level and could indicate saturation of the market.

The first and second place is maintained by Chile and the Philippines as their leading role as suppliers throughout the course of the years, although the carrageenan supplies in the U.S. were expanded by Indonesia and Malaysia. Certain reports indicate that the consumption of carrageenan in the United States is 60% reliant on imports as reported in certain reports.

Agar imports still stay steady from the United States, at about 1300–1500 tons each year. Countries like Chile, China, Morocco, and Spain continue to control commerce in this sector (Table 8.5).

Table 8.5. United States of America: Imports of Seaweed, 2013–2016

	Weight in Tonnes; Value in USD Thousands					
	2014		**2015**		**2016**	
	Weight	**Value**	**Weight**	**Value**	**Weight**	**Value**
Edible seaweed	7,180	60,670	10,695	72,628	8,560	55,883
Non-edible seaweed	18,030	46,698	14,826	36,400	20,959	38,481
Carra-geenan	11,475	114,695	11,192	105,699	9,959	82,830
Agar	14,17	34,379	1,565	38,500	1,383	31,949
Total	**38,102**	**256,442**	**38,278**	**253,227**	**40,861**	**209,143**

Sources: U.S. Department of Customs and Border Protection; U.S. Census Bureau.

8.7. MARKET TRENDS FOR DIFFERENT PRODUCT TYPES

In the international seaweed market, China has major grip on it and is famous for export of semi-finished products and as importer of raw materials in the last few years. Nevertheless, the emphasis is increasing on adding added value to the growing demands for carrageen and agar, especially from developing countries in other countries in Asia and Latin America. In the processing of food in Asia, Latin America, and the Middle East, the use of algae as a binder and thickener has substantially increased.

The large-scale importers of dried and semi-processed seaweed are found in Malaysia and Thailand in south East Asia. In addition, local demand for such products is increasing in Indonesian food processing industries. In the countries like India, Bangladesh or Pakistan markets in South Asia, where ready to eat food requirements is increasing; there is growing imports of seaweed items such as jelly, among other products.

In Argentina, Brazil, Mexico, and different other countries of continent of Latin America, it has been observed that there is an increase in number of hydrocolloids being imported for application in the food industry.

8.8. SEAWEED EXPORT TRENDS

There was a combined 30% share in the worldwide export volumes of the total reported sea algae exported by the top five suppliers, Indonesia, the EU, Chile, and the Philippines. However, China, the EU, Chile, and the Philippines were the leading suppliers in terms of export value owing to, amongst others, higher prices of carrageenan and agar.

The total international market supply of marine algae from these countries increased marginally in the 2012–2016 period apart from the Philippines.

The Indonesia exports dried seaweed and it comprises majority of its share of exports that was made up of a much lower export rate of dried algae (USD 800–USD 1000 per ton) in comparison to semi processed carrageenan (USD 8000–USD 10000 per ton) and agar (Table 8.6).

Table 8.6. World: Top Exporters of Seaweed, 2012–2016

	Weight in Tonnes				
	2012	**2013**	**2014**	**2015**	**2016**
Indonesia	174,836	182,446	206,572	212,245	188,760
European Union	82,171	95,411	103,447	101,546	101,594
Chile	77,990	88,409	81,265	75,238	82,482
China	69,239	64,284	68,135	70,828	78,681
Philippines	50,089	60,568	45,675	42,093	42,866
Total	**279,489**	**308,672**	**298,522**	**289,705**	**305,623**

Sources: FAO.

In 2016, the major producers in the international market were China, Indonesia, Japan, and the Republic of Korea, which accounts for around 75% of the volume of edible dried seaweed. Though, in terms of profit, the Republic of Korea became the biggest exporter exporting dried seaweed to the Japanese market, in particular Undaria or Hizikia, for the purpose of food for human consumption.

8.8.1. Indonesia

The Indonesian marine exports are overtaken by edible and non-edible raw dry goods, the major market for which is China. The countries like Chile, Malaysia, and the Philippines, that import dried seaweed for the processing of carrageenan and agar, are also major markets for those categories. In the 2012–2016 periods, the volume of exports of carrageenan has declined, with certain fluctuations, to about 5000 tonnes.

The EU, Japan, and the United States of America were the export markets for Indonesian carrageenans in that period. The exports to Australia, Brazil, and the Russian Federation have also grown significantly in recent years.

8.8.2. Chile

The major export commodity from Chile was non-edible dried seaweed mostly Gracilaria in context of volume and value, in 2016, around 80% of that product group went to China in the time period 2012–2016 this trend was continuing. In 2015, carrageen exports reached its peak, however decreased in 2016. The EU and the United States of America were the main markets. However, the exports to Argentina, Brazil, Ecuador, Mexico, and

Peru in Latin American markets have been increasing. Chilean agar exports have declined in the meantime (Table 8.7).

Table 8.7. Chile: Exports of Seaweed, 2014–2016

	Weight in Tonnes; Value in USD Thousands					
	2014		**2015**		**2016**	
	Weight	**Value**	**Weight**	**Value**	**Weight**	**Value**
Edible seaweed	3,522	8,663	1,107	4,162	748	4,617
Non-edible seaweed	71,152	13,4188	67,052	96,563	75,170	103,281
Agar	1,780	48,754	1,825	48,790	1,562	39,030
Carrageenan	4,811	72,475	5,254	73,548	5,002	59,622
Total	**81,265**	**264,080**	**75,238**	**223,063**	**82,482**	**206,550**

Sources: National Customs Service (Servicio Nacional de Aduana).

8.8.3. European Union (EU)

The most important exports from the EU are non-eatable dried seaweed, carrageenan, and agar. The main exporters of non-eatable seaweed are France and Ireland Germany, Portugal, and Spain produce smaller quantity. The key EU exporting nations are carrageenan, France, Germany, the Netherlands and Spain. Nevertheless, exports only increased from Spain from roughly 2010, but declined from other countries. In the case of agar, France, Germany, and Spain are the main exporters and the French agar, especially for its high quality, is most costly (Table 8.8).

Table 8.8. EU: Exports of Seaweed and Seaweed-Based Products, 2012–2016

	Weight in Tonnes				
	2012	**2013**	**2014**	**2015**	**2016**
Seaweed (edible)	3,733	4,691	5,041	5,219	4,607
Seaweed (not edible)	35,138	42,910	48,833	55,974	53,723
Agar	2,897	2,431	2,634	2,766	2,462
Carrageenan	40,403	45,379	46,939	37,587	40,802
Total	**82,171**	**95,411**	**103,447**	**101,546**	**101,594**

Sources: Eurostat.

8.8.4. China

In 2016, China, a leading manufacturer and exporter of processed seaweed, exported seaweed worth 500 million US dollars. The dried algae exports are comparatively small and have been made up of higher value wakame and nori, for immediate consumption, among other items. The major markets in the same year were Japan, the Russian Federation, and southeast Asia.

In 2016, Carrageenan became the primary portion of China's exports of marine algae and grew dramatically, hitting 360 million US dollars from 43,000 tons in 2012 to 56,600 tons in 2016. Asia, the Europe Union and the United States of America were the main export markets.

Exports of agar have rose from 4,000 tonnes, in 2012, to close to 6,000 tons in 2016, the main markets in the EU being Germany, Italy, Spain, and other countries like Indonesia, the Russian Federation and Thailand. Exports in the developing countries have also increased to numerous other markets.

8.8.5. The Philippines

The half-processed and processed seaweeds are part of the export of seaweed in the Philippines. The limited fraction of dried seaweed is also exported by the government.

The Philippines is China's second largest semi-refined and processed carrageenan provider or exporter in Asia. In the 2012–2016 periods, carrageenan exports increased about 33%, estimated in 2016 for USD 190 million. The principal carrageen market is the United States of America, followed by Brazil, the EU, Indonesia, Japan, Mexico, and Thailand. In addition, exports to Argentina, Australia, Brazil, Indonesia, and the Russian Federation have increased.

In some recent years, there has been a decrease in dry seaweed exports that are mainly of edible category. The volume of 11,000 tons in 2016 was significantly smaller.

8.9. PROCESSED SEAWEED IN INTERNATIONAL TRADE

There are typically three types of seaweed supplied on foreign markets:

- Ready to eat edible algae products;
- Serve as raw material for further processing of dry seaweed; and
- The seaweed turned into agriculture use chemicals like agar or

alginate and carrageenan. Market dynamics are largely attributed to the availability of hydrocolloids, including caragene and agar, commonly traded and used for food and non-food consumption in many countries.

8.10. PRICE TRENDS ON THE INTERNATIONAL MARKET

Overall, there is a differentiation in prices across different species, product types, consistency, and country of origin to mention a few. Price variations exist periodically as a result of supply and demand as well.

The main importer in the world, China decides the foreign price of carrageenophytes by Kappaphycus alvarezii and E. deliticulatum (E. spinosum) that are harvested in Asia and the Pacific region (Table 8.9).

Table 8.9. China: Average Import Price Trends for Edible Dried Seaweed (HS 121221) by Country of Origin, January–July 2015–2016

Exporting Country	**USD/Ton**		
	2015	**2016**	**2017**
Indonesia	1,039	716	916
Philippines	956	801	906
Malaysia	1,236	739	977
Japan	3,307	5,404	2,965
Chile	3,179	2,043	3,683
Taiwan Province of China	3,938	7,657	7,576

Sources: China Customs.

The cost of carrageenophytes is very subject to variability, particularly of E. spinosum. As demand rises, producers are raising production and generating an over-supply that then causes price declines sharply. Price is becoming extremely rapidly strong.

Farmers prefer to leave their farms while prices are low, thus reducing production and driving prices up again. In the case of processed seaweed, there are also many factors that affect different hydro-colloid prices like the quality, grade type, brand name, and country of origin and the features of physical products. The products from China, Indonesia, and the Philippines that are of Asian origin tend to be under the same spectrum of prices.

8.11. MARKET OUTLOOK

In view of its extensive use in the food and non-food sectors, global production of seaweed has risen significantly in the last decade. Market for fresh and preserved kinds, like dried seaweed, is mostly confines to eastern Asian markets and everywhere else in the world, including Japanese cuisine.

Even then, the effect on food and non-food world consumer markets has further been felt by hydrocolloid products that are carrageenan, agar, and alginic acid.

Carrageen is the most commonly produced and utilized of refined marine algae products. It is estimated that its current international trade is 120,000 tons. Although currently the EU and the United States are large markets, the requirement for carrageen, particularly in countries which are not manufacturing carrageenan, is increasing.

This pattern would definitely hit other developed countries. Carrageenan industry trends are favorable; in many emerging markets and business segments significant growth is predicted.

8.12. FOCUS ON THE HALAL MARKET

The international halal market is forecast to grow to USD 1.6 trillion by 2018 as predicted in a study carried out by the Chamber of Commerce and Industry of Dubai (DCCI, 2014). For that, 630 billion US dollars is given by the halal food industry alone to providing services. Catering for Muslims' religious responsibilities and desires has become a major organization, specifically in regard to the food, clothing, and tourism sectors, and this is very much visible in no other area than the Asia-Pacific region.

In order to sell their products and gain a market share in the halal food segment in Muslim countries, the countries like Japan, the Republic of Korea, the Philippines, Singapore, and Thailand are telling to their companies to obtain the widely recognized, all-important Halal certification for food products.

From Asia in the leading top, ten of the halal market the 8th and 9th place is occupied by only two countries Malaysia and Indonesia. Interestingly, the only Muslim countries (8th and 9th ranking) in the top ten in the world's leading halal markets are Malaysia and Indonesia.

The tourists from Muslim countries to Thailand from Asia and the Middle East have more than twice between 2006 (2.6 million) and 2016 (nearly 6 million) in Asia and in the Middle East as per the data from country government. Thailand has a strong domestic market within its large Muslim population in the southern part of the country, besides catering for tourists.

As Thailand is the ninth-largest exporter of halal food in the world, most of those products can be generally believed to be exported to other countries. In 2016, the Ministry of Industry of Thailand has estimated THB 180 million (USD 5.1 million) in a budget to support the promotion and exports of halal products so in near future this pattern will be visible in near future.

In order to make one of the top five halal food-exporting countries in the world by 2020, it is declared by Thailand got industry department that it will launch schemes for small and medium level business enterprise halal products on the basis of one tambon, one product. The National Food Institute of Thailand has assigned the amount of THB 120 billion (USD 3.4 million) and the Department of industrial promotion has allocated THB 60 billion (USD 1.7 billion), as well.

The halal-food products will get halal certificate in a hassle-free manner by the central Islamic Committee in Thailand and money has already given by govt. for this purpose to this group.

It is estimated that about 50% of the halal food exports comprise of rice from Thailand accompanied by sugar that is nearly 20% but up to what extent the seaweed and other algae comprises the remaining is not exactly known as per the national food institute.

There is, however, an agreement between the Institute and the Japan Halal Development Foundation where former Japanese owners will be supported by the Central Islamic Committee of Thailand in the purchase of the approved halal trademark. The outcome is that Japanese food, such as seaweed, will be available in the international halal market.

Singapore needs to be listed as it is an important illustration of business innovation. Singapore provides what several other countries may need halal consulting that is knowledge and understanding about the way Muslim countries are to produce, prepare, and sell products because the island state has no large land available for the development of massive food processing facilities. More than ten active halal consulting firms support companies in obtaining local and global halal certification.

In 2014, The Majlis Ugama Islam Singapore, the Halal certifying agency in Singapore, formed a company named Warees Halal Ltd. as its executive arm in 2014. The seminars conducted by Warees Halal Ltd were attended by companies from Japan and the Republic of Korea. Approximately 4 million SGD that is USD 2.95 million was collected from halal certification in Singapore that year as per the yearly report of Singapore's certifying body in 2014.

The companies seek plant-based alternatives for animal proteins and nutrients to meet Halal requirements. In addition to algae as a primary food supply, seaweed hydrocolloids are best adapted to supplement animal hydrocolloid gelatin, caseinate, whey protein, soy protein, white egg protein, and chitosan.

It should also be remembered that halal food follows the criteria for kosher food for Jewish populations as well as the US health guidelines. The Food and Drug Management and Codex Alimentarius Commission therefore acceptable to all, Muslims or others (Figure 8.4).

Figure 8.4. Halal certified meat in store in Malaysia.

Sources: Image by Wikimedia.

8.13. GLOBAL CERTIFICATION

In the developed world nations, the consumers are demanding greater proof that their manufacture and processing has been performed in a responsible fashion, from pollutant-free raw material to the final product that is safe to eat by health standards and same is true for products for halal market. Things that have been approved in this manner will be priced at higher value and kept in organic sections of large stores.

The qualification schemes like the Seaweed Standard of the Joint Marine Stewardship Council and Aquaculture Stewardship Council are being established in an answer to the need for internationally recognized criteria which will promote biodiversity and protection and are scheduled to be published during their introductory process in October 2017.

The primary process is available for farmed and wild seaweed operations that conform to the requirements like biodiversity, minimization of environmental effects, successful management, social responsibility, and group partnerships. During this process, the certificates issued will be valid for three years. The Seaweed Level complies with international ethical behavior, namely FAO Ecolabeling Standards for Fish and Fishery goods made by marine fisheries and ISEAL Good Practice Codes.

8.14. CONCLUSION

Seaweed has been used globally from ancient times for millennia at the very beginning only as a food supply for coastal populations. The global marine industry currently accounts for much more than 6 billion US dollars a year that is around 12 million tonnes, where about 85% are animal goods directly or indirectly utilized for human consumption.

The aquaculture market has increased massively as wild harvests reduced because of over-exploitation.

The overall world seaweed production was 30.4 million tons in 2015, and from the culture areas was 29.4 million tons. There is a need for ongoing research to detect rapidly growing, high-yielding, and disease-resistant species together with the need to make sure the long-term economic and environmental development of this sector by manufacturing countries.

The seaweed is not only viewed as a food source for populations at local level but also as an important source of socio-economic subsistence in terms of its culture and trade. The business of seaweed derivatives like carrageenan, agar, and alginates are an important source of trade income and make up nearly 40% of the hydrocolloid market in the world.

Therefore, the consistency of the supply of high-quality products is important and can only be accomplished through aquaculture.

There is very fast growth in international seaweed market, powered by the demand for goods in the categories of fit and not suitable for human consumption. The Seaweed and seaweed products were traded in 2016 for approximate 10.6 million US dollars.

Through an annual growth rate of approximately 10%, by 2025 the overall valuation could reach USD 26 million. Red varieties represent marginally more than half of the global value of red, brown, and green commercially important seaweed species. This pattern is anticipated to remain because of its growing use in the food industry both in Asia and worldwide, the most important growth center.

The Asia and the Pacific are by far the largest consumer market that is China, Japan, and the Republic of Korea, consuming about 60% of the supplies of seaweed and hydrocolloid, followed by Europe and the Americas.

REFERENCES

1. Dawande, R., (n.d.). Seaweed Market Size and Share | Seaweed Industry Analysis-2024. [online] Allied Market Research. Available at: https://www.alliedmarketresearch.com/seaweed-market (accessed on 10 July 2020).
2. Ferdouse, F., Løvstad, H. S., Smith, R., & Yang, Z., (2018). The Global Status of Seaweed Production, Trade, and Utilization (p. 3). [eBook] Food and Agriculture Organization of the United Nations. Available at: http://www.fao.org/3/CA1121EN/ca1121en.pdf (accessed on 10 July 2020).
3. Marketsandmarkets.com. (n.d.). Commercial Seaweeds Market | Size, Share, Trends, and Forecast to 2023 | COVID19 Impact on Commercial Seaweeds Market | Marketsandmarkets. [online] Available at: https://www.marketsandmarkets.com/Market-Reports/commercial-seaweed-market-152763701.html#:~:text=%5B139%20Pages%20Report%5D%20The%20commercial,is%20from%202019%20to%202023 (accessed on 10 July 2020).
4. Researchandmarkets.com. (n.d.). Seaweed Market: Global Industry Trends, Share, Size, Growth, Opportunity, and Forecast 2020–2025. [online] Available at: https://www.researchandmarkets.com/reports/5009064/seaweed-market-global-industry-trends-share?utm_source=dynamic&utm_medium=BW&utm_code=ggqjm7&utm_campaign=1375254+-+Seaweed+Market+Insights+Report%2c+2020-2025+Featuring+Profiles+of+Key+Players+Acadian+Seaplants%2c+Cargill%2c+DuPont+de+Nemours%2c+Irish+Seaweeds%2c+Leili%2c+Mara+Seaweeds%2c+and+GGOG&utm_exec=joca220bwd (accessed on 10 July 2020).
5. vegconomist-the vegan business magazine. (2019). Commercial Seaweed Market to Exceed USD $92 billion by 2025-Vegconomist: The Vegan Business Magazine. [online] Available at: https://vegconomist.com/market-and-trends/commercial-seaweed-market-to-exceed-usd-92-billion-by-2025/ (accessed on 10 July 2020).

INDEX

Figure 6.4. Anti-wrinkling of skin.

Sources: Image by Flickr.

6.5. CONCLUSION

From ancient times, people have been using marine food resources, and at present humans know that marine ecosystems contain a largely unexplored biodiversity, that presents significant potential for a variety of industrial sectors especially in the cosmetics sector. The seas and the oceans can supply the potential excipients and additives to be useful new extracts and molecules but above all active substances in the formulations of cosmetics that meet the requirements of so-called natural products attracted by consumers and the maritime world marketing concept. There is no doubt that moisturizing and anti-aging products are the product categories most affected.

REFERENCES

1. Couteau, C., & Coiffard, L., (2016). Seaweed application in cosmetics. *Seaweed in Health and Disease Prevention*, pp. 423–441.
2. Dhargalkar, V., & Pereira, N., (n.d.). *Seaweed: Promising Plant of the Millennium*. [eBook] p. 2. Available at: http://drs.nio.org/drs/bitstream/handle/2264/489/Sci_Cult_71_60.pdf?sequence=1 (accessed on 10 July 2020).
3. Harkulkar, G., Kumar, V., Gauri, S., & Suresh, S. P., (2012). *Medicinal Value of Seaweeds and its Applications: A Review*. [online]. ResearchGate. Available at: https://www.researchgate.net/publication/280527895_MEDICINAL_VALUE_OF_SEAWEEDS_AND_ITS_APPLICATIONS_-_A_REVIEW (accessed on 10 July 2020).
4. Jesumani, V., Du, H., Aslam, M., Pei, P., & Huang, N., (2019). Potential use of seaweed bioactive compounds in skincare: A review. *Marine Drugs, 17*(12), p. 688.
5. Pereira, L., (2018). Seaweeds as source of bioactive substances and skin care therapy: Cosmeceuticals, algotheraphy, and thalassotherapy. *Cosmetics*, 5(4), p. 68.
6. Wijesinghe, W., & Jeon, Y., (2011). Biological activities and potential cosmeceutical applications of bioactive components from brown seaweeds: A review. Phytochemistry Reviews, 10(3), pp. 431–443.